BEEKEEPING

Frank Vernon was born in Malvern (Worcs.) in 1910 of Anglo-French parentage, and was educated in England and France, graduating from Lille University in 1931. In 1945 he began beekeeping and soon became fascinated by the challenge of bees. From 1951 to 1970, while working as warden of a Youth Hostel, he was able to study beekeeping seriously at the Hampshire College of Agriculture at Sparsholt, under the guidance of that most dedicated of beekeepers, Capt E. J. Tredwell, and at the same time learned many tricks of the trade from local bee farmers. He became known to beekeepers throughout England for his colour slides and talks on matters relating to bees and beekeeping. In 1970 he became County Apiarist under Capt Tredwell at Sparsholt, and two years later transferred to the Ministry of Agriculture at Winchester as bee disease officer. For relaxation he looks after half a dozen small apiaries of his own and is currently engaged in cataloguing the Bee Research Association's Collection of Historic and Contemporary Beekeeping Material. He has a son and two grandchildren. He recently had a pacemaker implant in his heart and now feels as good as new.

TEACH YOURSELF BOOKS

BEEKEEPING

Frank Vernon

TEACH YOURSELF BOOKS
Hodder and Stoughton

First printed 1976

Copyright © 1976
Frank Vernon

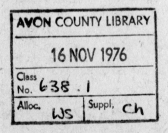
ISBN 0 340 20382 X

Printed in Great Britain
for Teach Yourself Books, Hodder and Stoughton, London,
by Hazell Watson & Viney Ltd, Aylesbury, Bucks

Contents

List of Plates

Foreword

Beekeeping is an age-old craft. The beekeeper's ability to handle bees, and to win a harvest of honey and wax from them, has always been a source of wonder to the layman. Throughout the world, beekeepers thus command a special respect.

What is new to the present generation of beekeepers is that they have the advantage of research work done in the last few decades. Some of these researches have enabled us to understand *why* bees behave as they do in specific circumstances. This has enriched the interest of beekeeping, and made it an even more attractive pursuit for people who enjoy observing and understanding wild life. For, like all insects, bees will always remain 'wild'; unlike farm animals, they cannot be domesticated. Moreover bees—ranging from purely solitary to highly social—are increasingly used in environmental education, and the whole range of flowering plants acquires a new interest when one considers each one as a source of nectar or pollen for a foraging bee.

As this book shows, one of the easiest forms of table honey to produce is honey in the comb, a food *par excellence* for those who prefer what is unprocessed by men or machines. The beekeeper with a few hives, for whom the book is written, may well prefer to pack his honey in jars, but even so it will not be subjected to much more handling.

Frank Vernon has written this book for newcomers to

beekeeping, and for those in their early years of experience with bees. He himself has kept bees throughout the period when so many exciting discoveries have been made about bees—about their chemistry, senses, means of communication, and so on. His instructions and explanations are written clearly and simply, but take into account this new knowledge. I hope that the book will encourage more people to keep bees and also increase the pleasure and profit enjoyed by those who already do so. It is not only the beekeepers who will benefit: every bee that goes foraging is a potential pollinator, and increased pollination can mean increased fruit and seed crops.

December 1974

Evan Crane,
Director,
Bee Research Association

Preface

Whoever sets out to produce a book on beekeeping travels a well-trodden path. It does become necessary from time to time, however, to start again from basic principles, in the light of a changed environment and new technology. In this short book, I have tried to include all that a newcomer to the craft need know to manage bees successfully. For those who wish to push their knowledge further, the book has consciously taken into consideration the syllabus of the British Beekeepers Association Preliminary Examination. This has led me to spend time at each step explaining the reasons 'why', rather than be content with merely setting out directions for 'how' to proceed. In this way I hope to have improved the student's ability to grasp and remember the practical advice. The theory should provide the necessary background to allow the reader to vary the instructions to suit individual needs and idiosyncracies, as well as help those who keep bees in other climates to find appropriate solutions to local problems.

Even such a small work as this requires the co-operation of many people. I would like to thank my wife for her constant assistance and support, Mr and Mrs Cossburn for their advice and technical help, and the many beekeepers who came to my rescue during a serious illness which delayed the production of the book. Above all I would like to thank the Bee Research Association and its Director, Dr Crane, for their unfailing help and advice, both personally and through their many admirable publications.

Southampton
December 1974

I

The Bee Colony

The honeybee (*apis mellifera*) belongs to a large order of insects known as the *hymenoptera*, which includes ants, wasps and bees. Bees evolved millions of years ago from wasps (though not the kind that are so troublesome in the autumn). There are several kinds of bees in existence today:

Solitary bees, which have no form of family or social life.

Semi-social bees, which live a social life in summer but whose society disintegrates in late autumn. Mated females then hibernate and restart a new colony the following year. Bumble bees belong to this group.

Social bees, which have learned to store food against the season when no flowers are in bloom. Their societies are maintained permanently. This group includes the tropical stingless bees, *trigona* and *melipona*, the tiny Indian *apis florea*, the large *apis dorsata* and, of course, our honeybees.

From earliest recorded times, man has been associated with bees, eating their brood or honey, using their wax, or intoxicating himself with mead fermented from their honey. For centuries man kept bees in primitive hives built to resemble the natural cavities where he found them wild. In Europe, early beekeepers put their swarms into a clay pot or a basket woven from cane or withies. Later the basket was made by stitching together coils of straw rope. Most people have heard of this kind of hive or skep (Fig 1).

The modern hive allows us to open a hive, comb by comb, and to find out what goes on inside. As a result the technique of beekeeping has improved immeasurably in the last 100 years, in spite of the reluctance of beekeepers

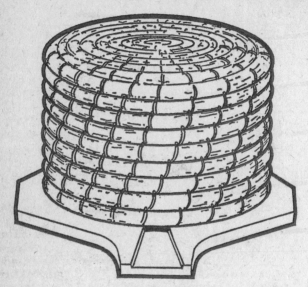

FIG. 1 The straw skip hive

to keep up with new discoveries. Today, bees are kept by amateurs and commercial and research establishments, and although these three groups have different aims, pleasure, profit and learning respectively, each group shares all the benefits to some degree.

The actual bees in a hive are collectively called a *colony*. The box or basket in which the colony is kept is the *hive*. Inside the hive the bees build a series of vertical slabs of wax combs; an examination of the comb reveals that they are made of horizontal and opposed hexagonal cells. The architecture of the comb can best be explained by the

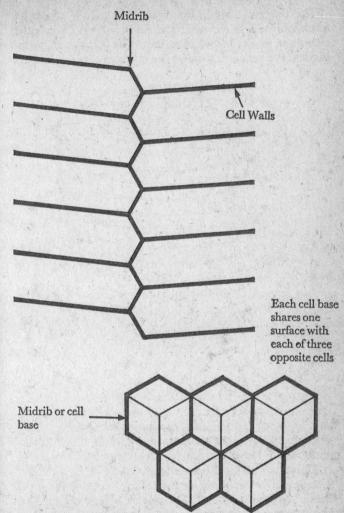

Midrib

Cell Walls

Each cell base shares one surface with each of three opposite cells

Midrib or cell base

FIG. 2   Comb construction

drawing in Fig. 2. Although the wax in the cells is very thin, the structure gives it such strength that a piece of comb 10 cm x 10 cm holds a pound of honey. The cells are used to raise the young brood, or to store pollen and honey for use at times when no food can be brought in from outside.

If you examine a comb from the centre of a hive, you will see that it is made up of two different sizes of cells: worker cells approximately 5 mm in diameter and drone cells over 6 mm. But at certain times of the year, you will notice cells standing out from the comb and hanging vertically downwards. These are queen cells. Their size varies from 2.5 to 4 cm in length and they taper down from 12 mm at the top to 8 mm at the lower end.

The brood cells, worker or drone, occupy the centre of the comb. An arch of cells containing pollen surrounds the brood, and further out still lies the store of honey. This internal pattern, although it varies in size, remains fairly constant throughout the year, but occasionally the pollen arch can be overrun by the eggs; the queen, however, never lays beyond the honey. The beekeeper must accustom himslf to recognise at a glance the appearance of brood, pollen and honey cells and the beginner should seek the help of a competent beekeeper to point them out to him or her.

All insect societies, including honeybees, have two basic characteristics in common. First, the society is centred round a female who is the progenitor of the whole colony. In that sense, it is a family. From this we must deduce that all the characteristics and instincts of the colony derive from the genetic material of the mother and the male with whom she has been mated. In a colony of bees, if we change the queen, we effectively change the colony. Secondly, insect societies are bound together by *pheromones*. Pheromones are chemicals which pass from one individual to another and effect the behaviour of the recipient. The study of pheromones is still at an early stage, and practical beekeeping may yet be revolutionised when more information is discovered about the nature and effects of these substances.

The bee colony is made up of three basic types of individuals, or castes:

The queen: technically the mother of the entire colony.

Drones: potential sires of the next generation of queens and workers.

Workers: sterile females who carry out all the social duties, except egg laying and insemination.

Life in the hive involves all the inmates continuously. The drive in the struggle for survival comes from the inexorable power of instinct. Insects do not have intelligence, as we and other mammals do. Instead, their activities are governed by behaviour patterns which are not learned but inherent in their genetic make-up, by controls in their glandular development, and by the distribution of pheromones.

One of the ways in which these pheromones are spread is the process of food sharing—*trophallaxis*. It ensures not only that no individual bee can die of starvation whilst others live in plenty (how unhuman bees are!) but also that the shared food acts as a vehicle for chemical messages. Thus each bee receives information about whether, and what kind of, food is coming in; whether the queen is present and laying; whether brood is present; and probably very much more.

The instincts of bees are so complex, and adaptable to such varied environmental and accidental circumstances, that many people have ascribed to bees intellectual powers which they just do not and cannot possess. In fact, if we hope to keep bees successfully, we have to work *with* their instincts. We have to understand what is going to result from any interference of ours and act accordingly. After all, we are the ones who are supposed to have intelligence.

All the bees in the colony have a number of features in

common, although the details may differ in the three castes.

Through the process of metamorphosis, the egg develops by stages to larva and pupa, and finally emerges from the cell as an adult bee. Three distinctly separate parts form the body of the adult bee: head, thorax and abdomen. A hard external skeleton of *chitin* encloses all the internal organs. This skeleton is built up of a series of overlapping rings or *segments*, easily discernible in the abdomen but hard to distinguish in the head and thorax, because there the segments have coalesced in the course of evolution. The abdominal segments can be seen to expand and contract as the bee breathes, and this movement is made possible by a soft, folded membrane connecting the hard chitin of two adjoining segments. Air enters and leaves through an aperture or *spiracle* on either side of each segment of the abdomen and thorax and is carried to the organs of the body via a system of tubes—*tracheae*—into air-sacs and thence by further tracheae branching out into minute tubes—*tracheolae*—to reach the remotest organs and bring them a supply of air. Unlike mammals, bee's blood carries around food but no oxygen; it flows loose in the body without either arteries or veins, pumped through an open-ended heart—the dorsal vessel.

Besides the digestive system of gut, crop, large and small intestines, and a rectum which stores exreta until the weather favours flights, the abdomen also encloses the reproductive organs: ovaries of the queen and workers, testes and penis of the drones. The two female castes possess a sting, connected to poison glands. Worker bees drive home this fact with unforgettable insistence, but queens reserve their stings for rival queens only.

The thorax bears on its underside three pairs of articulated legs and two pairs of wings hinged to the sides. The forelegs have a neat antenna cleaning brush. While at rest, the bee folds its wings over its abdomen, but when preparing for flight it slides the forewing over the hind wing. A channel under the trailing edge of the forewing locks into a series of hooks above the leading edge of the hindwing, effectively turning the two wings into one. Two pairs of powerful muscles inside the thorax alternately depress and raise its domed cover, transmitting the movement by a lever and fulcrum mechanism to the wings themselves, thus enabling the bee to fly.

The senses

Two large compound eyes surround the sides of the head, and their many facets or lenses give the bee a wide field of view, although each facet covers a very narrow angle of about $1\frac{1}{2}$ degrees. Bees can distinguish light, shade, shape and movement, although the patterns they recognise are very different from those we normally consider as distinct. They see a similar range of colours to our own but with a shift away from the red end of the spectrum—bees see red as black—towards and into the ultra violet, which is beyond our perception and appears black to us. Bees' 'colour circle' includes two secondary colours, yellow with ultra violet and blue-green with ultra violet. In addition, their eyes allow them to 'read' the angle of polarised light from the sun or blue sky and to 'calculate' and correct for wind drift during flight.

Three simple eyes or *ocelli* lie grouped in a triangle between the compound eyes. It would appear that these ocelli act as incitors to the other eyes, increasing or decreasing their light receptivity according to circumstances.

The two antennae on the head are smell and touch organs, which can be seen to move with greater flexibility than any other part of a bees' anatomy, thanks to their very fine segmentation—twelve segments in the females, thirteen in the males. Plate-like sensors on the antennae perceive smells, and hair-type receptors transmit a sense of touch. Similar touch sensors are distributed about the body. It is worth noting that bees can smell water, but in other respects their sense of smell is not very dissimilar to our own. There are also many other sensors on the antennae.

Taste 'buds' can be found in profusion on the mouth parts and on the lower joints of the feet, a fact which may surprise many readers, but this location is fairly common in the insect world. Their taste extends, as does ours, to sweet, sharp, salt and bitter. The mouth parts, which are most concerned with taste, include a pair of smooth, hollow-grooved *mandibles* or jaws and a compound proboscis which lies folded back under the head when not in use. For feeding or gathering nectar, the proboscis is extended, and its four overlapping curved elements form a tube surrounding the tongue or *glossa*. Food can only be sucked in when a V-shaped notch, placed where the tube folds, is tightly gripped by the mandibles. A suction pump in the head, coupled with a licking and up and down movement of the glossa, enables the bee to ingest fluids, even when they are as viscous as honey.

A bee's sense of hearing differs entirely from our own. Very simple hearing organs inside the hind legs allow them to hear low-frequency sounds transmitted through solids but not sounds transmitted through air, since they have no tympanum which could allow them to receive air-borne vibrations. However, if their feet are in contact with a vibrating solid, they are well able to perceive the kind of 'sound' produced.

Bees have other senses too, although their location is little known. For instance, an exact temperature sense enables them to maintain the brood nest at a constant temperature and to regulate their winter behaviour according to a precise pattern determined by the prevailing weather conditions. They can also sense the viscosity of the honey they store and very seldom make the mistake of sealing syrups dilute enough to ferment.

Much more could be said, but the more advanced bee-keeper must be referred to the specialist works listed at the end of the book. This chapter has concentrated mainly on the features common to all the bees in the colony. The next will attempt to bring out the differences between them and the way the three castes conduct and organise their social life.

2

The Three Castes and Social Life

It must always be borne in mind that the unit of life in a hive is the whole bee colony and that the individuals can survive separately for very short periods only. When we study them one by one, we must also consider their inter-relationships.

The queen

A colony has only one queen (although, just occasionally, an old queen can be found laying side by side with her daughter in the same hive). Her legs bear no built-in tools for pollen gathering, nor does she possess food glands to feed her numerous progeny. Her thorax measures about 1·5 mm more in diameter than a worker's and she appears to be somewhat longer, particularly when she is in full lay and her abdomen is distended by two large ovaries which fill the abdominal cavity. The ovaries discharge eggs into two branch channels, converging into a single oviduct. As the eggs descend the oviduct, they pass the opening of another duct from the *spermatheca*, a small, round sac containing the sperm from the drone or drones, collected and stored at mating time. A gland on the spermatheca controls the discharge of a very few spermatozoa at a time onto each egg, just prior to laying, and thus ensures fertilisation.

The fertilised egg passes down onto a groove in the top of the queen's curved sting and is deposited, upright, on the cell base, where it will develop into a worker bee. If, however, the egg is not fertilised, the queen deposits it in a larger drone cell, where it develops into a male. Development of a new individual from an unfertilised egg is called *parthenogenesis*—derived from two Greek words meaning virgin birth; the queen is not a virgin, of course, but the male sperm plays no part in the production of drones. Male bees inherit only the genes of their mother and, genetically speaking, are just as much her twin brothers as they are her sons.

The queen's eggs can develop into drones, workers or new queens. Parthenogenesis accounts for the production of drones but does not explain how similar fertilised eggs can produce such different individuals as a queen and a worker. Surprisingly, the differentiation arises only through the variations in diet received by the larvae after they have hatched.

When conditions dictate to the workers that a new queen is needed, she is induced to lay a fertile egg in a large cell hanging vertically down from the comb; or, sometimes, a worker cell containing an egg or larva is enlarged downwards, at right angles to its original direction. A larva hatches from the egg three days after laying and at once receives an abundance of rich bee milk—sometimes called royal jelly—for a total period of five days. By this time the larva has grown to fill the base of the queen cell and worker bees cap and seal it. Eight days later still (sixteen days from the laying of the egg) a new queen will emerge. Except for some sips of honey while she is still a virgin, the queen will be fed royal jelly for the rest of her life.

By contrast, the fertile eggs laid in worker cells have another destiny. They hatch in three days and are fed the same bee milk as queens, but in much smaller quantity, for the first three days. For the next two days, the quality of their ration changes by dilution with honey and, perhaps, pollen. Worker cells are sealed over on the eighth day after laying, and the workers emerge a fortnight later, twenty-one days from the egg. Not only does the worker-type female differ from the queen anatomically, she also emerges equipped with a different set of inherited instincts. How can just two days of different feeding alter the very coding of the genes? There can be no question of choice of eggs. Any worker egg can be transferred into a queen cell and be made to develop into a queen and *vice versa*, always provided that the transplant is effected during the first two days of larval life. It is thought that the special food triggers off a different code, but science has yet to solve this most intriguing problem.

The virgin queen usually emerges into a colony from which the old queen has already flown. She starts her adult life by searching the combs for any potential rivals, either still in their cells or recently emerged. Those she finds she stings to death through the walls of the encasing cell. It would ease the beekeeper's life no end if she were more thorough in despatching her rivals, but alas, for reasons not fully understood, she seldom achieves complete elimination of all challengers. When two emerged virgins meet up on the combs, a battle royal ensues and one of the combatants dies.

Mating of the new queen must take place in flight and, for the good of the colony, as soon as possible, so that egg-laying can resume for replacement of the working population. The weather has to be bright and warm and it is usual for the nuptial flight to take place between four and nine days after emergence. The first few days are occupied in feeding and making reconnaissance flights to mark the exact location of the hive entrance. On the chosen day she flies towards a nearby congregation of drones. Copulation may take place with more than one drone on the same flight, or sometimes more than one mating flight may be made. When the now mated queen returns to the hive, she carries in her vagina the remnants of the last drone's sexual organs—the so-called mating sign. The sperms are stored away in her spermatheca, and after a day or two the new queen starts her life of egg laying. Estimates of the number of eggs she can lay vary enormously. In one or two days she may lay as many as 3000, but during the two or three weeks of her early summer peak the average daily output is unlikely to exceed 1500 to 2000. At other times, it is much less.

It has become known in recent years that the queen's function in the hive extends beyond mere egg-laying. In glands connected with her mandibles, she secretes a phero-mone which, for want of a better name, has been called *queen substance*. This has actually turned out, since Dr C. G. Butler discovered it, to be quite a number of sub-stances. The queen can often be seen on a comb sur-rounded by a group of workers, facing in towards her, as if they were courtiers round their sovereign. This 'court behaviour' of workers grooming the queen serves the essen-tial function of distributing the pheromones, at first to the 'courtiers' and then, by food exchange, to the other mem-bers of the colony.

The effects of queen substance in the colony are far reaching and can best be negatively described by enumer-ating the disruptions in behaviour which follow a cut-off in supply, as for instance when a queen is killed or re-moved. At first the bees show disquiet and search the hive both inside and outside for their missing queen. After a few hours, they start building queen cells over existing worker larvae, still young enough to be converted into queens, and floating them out of the old worker cell on an abundance of royal jelly into hastily built 'emergency cells'; but should some accident destroy them and the colony remain queenless and broodless for a few weeks, then some of the younger workers develop ovaries and start laying eggs. These laying workers cannot mate and their unfertilised eggs always develop into males, even when laid in worker cells.

Drones

Drones are male bees and cannot possess a sting. Their head has larger compound eyes than either of the female castes, their antennae have one segment more, their proboscis is short, and their thorax is even larger than the queen's. Their abdomen is wide and blunt, with a fringe of feathery hairs near the tip.

The unfertilised egg which develops into a drone is laid in a bigger cell than a fertilised egg, horizontal on the comb; but since the male is not only stouter but also longer than the worker, the cell it grows in must somehow be made longer than worker cells on the same comb face. When the time comes for the cell to be capped over, a raised dome is built over the larva to accomodate the extra length. The emergence of an adult drone occurs twenty-three to twenty-four days after the egg is first laid, but he is not potent and able to mate for another two to three weeks. After a few days in the hive he will undertake some play flights in fine weather to locate his home and then he flies off to join other drones in congregation areas somewhere in the vicinity. Drones from many hives and even several apiaries form the congregation. Well-defined flight lanes appear to be scent-marked, so that when a virgin emerges on her nuptial flight she readily finds her way to the males and successful mating ensues.

The colony starts rearing drones in early spring and maintains a certain number of them until the end of summer, just so long, in fact, as their services might be required in case of mishap to the queen. At the end of the honey flow, when the stores are being sealed over for the winter, the drones are disposed of. The hive guards just do not re-admit them to the hive. Those inside are driven out to starve, and in the space of less than a week the slaughter is over. There will be no more drones until next spring. If, by any chance, drones are maintained abnormally late, there is a good reason for it. Quite likely, an examination will reveal that the queen is being replaced, and the drones have been kept to ensure safe mating of a late queen.

Workers

From a fertilised egg can develop a queen anatomically fitted solely for egg laying, relying on others to provide her with food and shelter, physically resembling most closely that group of bees which live a parasitic life; on the other hand, the same eggs, fed differently after hatching, can produce worker bees which have inherited from their ancestors all the external and internal adaptations necessary for their survival. Their ovaries may be atrophied, but their bodies are beautifully organised for the many functions they are called on to perform during their short lives. It is not the purpose of this book to present a complete account of the anatomy and physiology of the honeybee, but without some understanding of the essential features of both it is unlikely that success in beekeeping can be attained.

Most people know one thing for certain about worker bees: they are armed with a poison sting which they use for hive defence. Barbs on the tip of the sting make it difficult for the bee to withdraw it from some kinds of skin, and in such cases the sting, poison sacs and a nerve centre are torn away from the bee and left behind in the victim. For some seconds after the bee has flown away the sting continues to drive down deeper into the skin, pumping in poison; the bee, minus the last segment of its abdomen, slowly loses blood through the wound and dies within a day or two.

The worker starts life as a fertilised egg in a worker cell. In three days, the egg skin splits around the middle, dissolving towards either end and a tiny larva hatches out. Nurse bees place a supply of milk near and around the larva for its nourishment, and after the third day the milk

'poison sac'

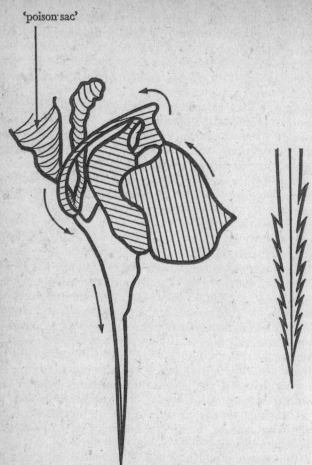

FIG. 3   How the sting works. Muscles pull the large quadrate plates. The arrows show how the movement is transmitted to each side of the sting in turn. Barbs hold one side, while the other drives deeper. Two sets of plates operate the lancets alternately.

is mixed with honey. After five days of feeding, the larva has grown until it completely fills the diameter of its cell base. It must, of necessity, uncoil itself and stretch full length in its cell. Adult bees then cap the cell over with a porous wax cover, slightly domed, while the larva undergoes a series of metamorphoses and moults, which transform it, stage by stage, into a pupa and finally an adult worker, over a period of thirteen days. Twenty to twenty-one days after the laying of the egg, the young bee gnaws away its capping and emerges, not without some struggling, into the milling, teeming life of the bee colony.

Within a few minutes, the young worker will start cleaning, stropping and polishing recently vacated cells, ready for the queen to re-use. House duties have begun, and for the next nine to fifteen days a fairly consistent sequence of such duties has to be undertaken. Cleaning up cells in which pollen had been stored leads the young bee to eat pollen; this makes up for two weeks of protein starvation during metamorphosis. The bee's consumption activates food-producing glands in its head and induces secretion of milk. Our worker has become a nurse bee and sees to the needs of the unsealed larvae, first attending the younger grubs and later the older ones, which require an admixture of honey in the food. And so it is led, naturally, to some contact with the honey stores or to nectar brought in by foragers. The nursing function belongs only to the worker bees and the queen is denied that whole area of motherhood. Who is the real mother in a hive? The queen who lays the egg, or the worker who feeds brood and raises the young?

Consumption of honey instead of pollen reduces milk production, and the young worker starts to relieve incoming foragers of their loads of pollen and nectar, storing them in the cells. Consumption of sugar increases sufficiently to activate four pairs of wax glands under the abdomen. Numbers of workers at this stage hang together in strings or chains and generate heat. Under these con-

ditions, wax oozes from the wax glands between the segments and, on contact with the air, dries into tiny wax
scales. The bees manipulate these scales with their mandibles and use them in building or repairing the combs,
or in starting up an entirely new home when they swarm.

Comb repairing involves a number of workers in clearing
up dirt and debris and throwing out dead bees. The hive
is always kept scrupulously clean, whenever the weather
permits flight. Journeys to and from the entrance lead the
young workers to try their wings on the hive front and to
begin ventilation duties. Ventilation plays a most important part in hive economy, fulfilling a number of vital
functions: gas exchange, temperature control in hot
weather, processing honey, and scent marking the hive for
young bees out on their location flights. A worker engaged
in ventilation stands at the entrance of the hive, facing
inwards, head down, with the tip of its raised abdomen
turned down to expose its scent gland. Several bees in this
position fan with their wings and create a draught *out* of
the hive and across their scent glands. The native bees of
Eastern Asia, however, reverse the procedure: they face
outwards and create an in-flowing draught. The scent
appears to be individual to each colony and can be picked
up several feet away by incoming bees. In very hot weather
these fanners keep the draught moving across the combs,
evaporating moisture from incoming nectar and so maintaining a constant temperature of $31°C$ to $33°C$, required
for steady development of the brood. At the same time,
exhaled carbon dioxide and water is driven out.

Guard duties follow or are combined with fanning.
These entail patrolling the entrance to the hive and chal-
lenging incoming bees for their 'credentials'. A bee chal-
lenged by a guard responds by adopting a submissive
attitude, curling over on its side and offering food. Any
bee, wherever it may come from, will be admitted if it is
bringing in food and submissive. If, on the other hand,
the incomer tries to dodge the guard and has nothing to

offer, it will be chased away and, if it persists, will be attacked and often stung to death by the guard. Occasionally, a large force of robbers can overcome the guards of a weak hive and steal all its stores of food. This kind of disaster can usually be blamed on the beekeeper's carelessness, rather than on the wickedness of his bees.

Cleaning, fanning and guarding have brought the worker to the entrance of the hive and soon lead it to making its first location or 'play' flights. On any fine summer day, a hundred or more bees can be seen facing homewards and bobbing up and down in front of the hive, from a foot to several yards away from the entrance. They are learning the exact location of their home, its position and setting, in relation to other features around it. The reconnaissance completed, foraging can begin. Generally, the worker does not just go out on its own and cast around in the hope of meeting up with a flower or two. Instead, it is recruited by other foragers to exploit known food sources. The gathering area encompasses a circle of about half a mile radius around the hive, but under exceptional circumstances this can be extended to 2 or more miles.

Nearly always, the forager starts by bringing in pollen, using for this purpose a number of tools distributed over its body. Entering a flower, it rolls over the stamens, or scrabbles with its legs among them, collecting the pollen in its body hairs. A bee's hairs grow characteristically twisted, forked or branching and are ideal for picking up the rough-skinned pollen grains. When the bee lifts off from a flower, it sweeps itself clean with its legs, all of which are fringed with stiffish brushes. The hind legs are provided with a neat pollen brush, on their inside faces for collecting up the pollen from the other legs. On the leg segment above the brush a stiff comb protrudes inwards, and a quick double sweep of each comb, across the brush on the opposite leg, rakes the pollen into a tiny pellet under the comb. A click of the lower segment drives this pellet up an inclined plane to the outside of the upper segment, into the *corbicula* or pollen basket, where successive pellets are retained by a single stiff central bristle and an outer fringe of long hairs. Moistened with nectar and patted into shape, the pollen loads are carried back to the hive and conveyed to a cell where they are discharged. A younger house bee rams the pollen down, while our forager moves to the comb face and recruits more bees to work the food source.

Recruiting foragers may seem easy to us humans, particularly those of us who spent some years in the Army, but in the insect world bees have achieved a remarkable evolutionary advance. A returning forager, bringing in either honey or pollen, proceeds to some spot on a vertical comb face and performs a pattern of movements, surrounded by an audience of apparently attentive bees. Professor von Frisch, in Austria, analysed these 'dances' and found that the dancers were able to impart to the onlookers information about the whereabouts of the food source, its direction, its distance and its substance. A bee performing a 'round dance' starts at a point and circles until it reaches the same point again. It then turns round, repeating the circle in the opposite direction. This dance indicates that forage is available within 10 metres or so of the hive. However, if the food lies beyond this distance, the round dance gives way to a figure of eight dance which includes a straight run across the circle, always in the same direction, in the course of which the dancer waggles its abdomen, hence the name 'waggle dance' given to this performance. By relating the angle of the straight run to the vertical on the comb and gauging the frequency of the waggle runs, von Frisch shows how the information imparted by the dance allows the recruited bees to leave the hive and find their pasture without further guidance. Four different dances have so far been recognised.

After gathering pollen for a few days, our now experienced bee turns its attention to foraging for nectar. With

Round dance.
Food at 10 m
No direction

Sickle dance.
Food at
about 50 m
In direction
of arrow

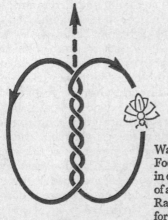

Waggle dance.
Food beyond 100 m
in direction
of arrow
Rapid circling
for nearby food –
slow for distant

FIG. 4   Von Frisch's bee dances

its long proboscis it explores the floral nectaries, licking and sucking up the dilute sugars exuded by the flowers. The bee adds the enzyme *invertase*—derived from glands in the thorax—to the nectar on its way into the crop, and this sets up a chemical change in the sugars which uses up part of the water in the dilution. On its return to the hive, the forager disgorges its crop contents to some of the house bees, who elaborate the nectar into honey, subjecting it to the draught in the hive so as to evaporate further water. When the concentration of sugars has reached somewhere near 80%, it is placed in the cells. At or near 84%, when yeasts can no longer ferment it, the honey is sealed over with a flat wax capping.

Those foragers who survive an exhausting period of nectar gathering spend their last few days collecting gums and resins from leaf buds and using them to fill up small cracks and cavities in the hive. When so used inside the hive, these sticky substances are called *propolis*. Until recently, propolis was considered useless to man, but there is now some evidence to show that it contains valuable antibiotic properties and supplies are being sought by several firms, who are prepared to pay around £45 a kilogram for clean samples. Some of the older bees gather water, either from dew on the leaves in early morning or from damp earth or pond edges. Water is always needed in the spring to dilute stored honey for consumption. In spite of the fact that they take such pains to concentrate honey for storage, bees are unable to consume it until it has been restored to its original thin consistency. In arid, hot, desert areas, dew is collected, and suspended droplets are evaporated during the heat of the day to ensure that the wax comb does not melt.

The reader may well be wondering by now just how long a bee lives. Unfortunately, the question betrays its human origin. Bees do not live and die by the clock and calendar, as we incline to do. A queen, for instance, lives until she is superseded by a new one, after or just before she has outlived her fertility, and that may mean a few months, a year, two years or even more. I have known one to be replaced during her sixth year. It appears likely that most queens are killed off and do not just die. Drones live until they mate or until the end of the main honey flow, but they also do not die, they are slaughtered. According to the time they are born, they may live months, weeks or only days, but their demise is always synchronised with the end of summer. Workers, on the other hand, die. Their life peters out as soon as they have completed the series of functions described above, but the time lapse in

Number of days required for development of the three castes

Stages	Worker	Queen	Drone
Open cell:			
Egg	3	3	3
Larva	5	5	7
(4 moults)			
Sealed cell	12–13	8	13–14
(2 moults)			
Total days to emergence	20–21	16	23–24

weeks or months has no significance. A worker emerging in spring passes rapidly through the exhausting phases of nursing, wax production and foraging in the space of four to six weeks, and then dies. But as summer draws to a close, brood rearing, foraging and other activities are curtailed, and many of the young bees find no outlet for their secretions and functions. In these circumstances, in a healthy colony, the workers stay alive and do not age, at least not until the following spring, when activities resume again and they carry on where they left off. They expend themselves to raise and provide for the new season's brood, and die when they become exhausted.

Such, then, are the principal features and activities of the members of a bee society. By knowing them, we shall be able to understand our bees and guide their instinctive reactions into paths profitable both to them and to ourselves. Much has been omitted, much is still to be unravelled, but enough should now be available to form a basis for our practical beekeeping.

3

The Hive

The modern hive is a set of simple superimposed boxes, in which bees are kept on moveable combs. The facility to move combs and add or remove boxes depends on the combs being enclosed inside hanging wooden frames, separated from the box and from each other by a *bee-space*.

Bee-space

Bees normally fill with comb all the space available to them in a hive. In plain boxes or skeps, they build natural comb fixed to the roof and sides. Under such conditions, any management which involves interchanging or moving combs is quite out of the question. In 1851, L. L. Langstroth in the USA discovered that if the combs were fitted into wooden frames, separated from the box sides by a gap of 5 to 9 mm, the bees would leave this space open and free from wax or propolis, in the same way as they left an open space between comb faces to enable them to have access to the brood and stores in the cells. The 5- to 9-mm gap in a hive is called a bee-space.

If the clearance falls below 5 mm it becomes filled with propolis, presumably to prevent invasion by small predators and to keep the bees' home watertight. Spaces larger than 9 mm get filled with *brace comb* at right angles to the general comb direction. Brace comb is built in regular hexagon cell pattern. Where the area they have to fill is

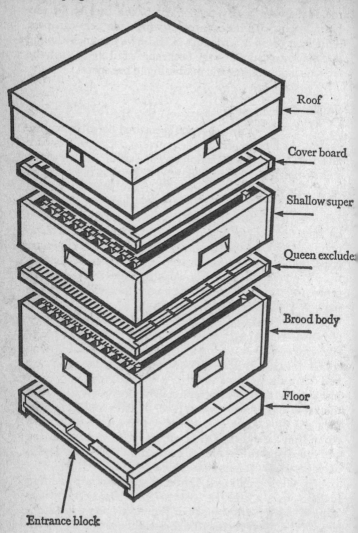

Roof

Cover board

Shallow super

Queen excluder

Brood body

Floor

Entrance block

FIG. 5 A modern hive, the 'Langstroth'

too shallow to justify proper comb, the bees build wax burrs, often with holes through which the bees can pass. This *burr comb* is a product of bad hive design, although it can also be caused by the beekeeper's failure to push the frames close together and maintain the bee-space.

Frames

As we have seen, the bee-space is ensured by enclosing the combs in a hanging frame, designed to permit a bee-space all round it in every direction except a small area on which it is suspended. The top bar extends beyond the frame to form 'lugs' which drop into a rebate in the hive wall and position the frame centrally. This keeps the bee-space between the frame sides and the hive. Some device has to be incorporated or added to the frames to space them regularly and accurately across the box. Thus frames are either self-spacing or use 'spacers' of wood, metal or plastic. Brood combs are often spaced at 37 mm centre to centre, but there is a tendency now to prefer the more natural 35-mm spacing.

Spacers are only useful if the bees build straight slabs of comb in the exact centre of the frames. We can take advantage of the fact that bees work at comb building collectively. Each bee takes up the work as left by others and carries it forward as far as it can with the wax it has just produced. It is therefore possible for us to do part of the bees' work ourselves and leave them to finish the complex task of making the deep cells. The wax mid-rib, with part of the cell walls, can be embossed on a sheet of wax and fitted into the frames, supported by thin wires. Sheets of embossed wax are known as *foundation*.

A hive can be enlarged during the honey flow by adding further boxes above the lowest. Upper boxes, known as 'supers', have to allow for their frames to hang a bee-space clear of the frames below. To this end the boxes are invariably made 8 mm deeper than the frames. The bee-space so

made can be left above the frames, as in the internation-
ally used American type hives, or it can be left below the
frames, as it is in the British type of hive. For obvious
reasons, a box with lower bee-space cannot be placed above
the other kind, or vice versa.

Hive components

Modern supering hives consist of a number of well defined
and purpose-built parts.

A floor: a flat, three-sided tray, closed or opened on the
fourth side by a block of wood which has a 75 mm × 8 mm
slot cut across one side to give restricted entrance when
required.

A brood chamber: made to contain the frames of comb
which form the permanent bees' nest, together with their
stores of pollen and honey. Ten or eleven frames, spaced
35 or 37 mm apart, furnish the brood chamber.

Supers: hive boxes to receive the honey are usually
about half the depth of the brood chamber, hence their
alternative name of 'shallows'. They are also lighter to
lift and manipulate.

An inner cover or *crown board* covers the topmost box.
It is usually made with two so-called 'feed holes' which
are cut to allow the bees to rise into a feeder or to take the
one-way bee traps that clear bees from the supers. In this
way the crown board is made to double up temporarily,
as a 'clearer board'.

A roof keeps out rain, protects the sides and should
allow for adequate ventilation.

Appliance manufacturers make hives with great pre-
cision. For British hives they work to BSI specifications
(BS 1300 1960). The beginner who wishes to make his
own hives would be well advised to borrow good, sound,
recently made models from another beekeeper and copy
them exactly. Full constructional details for four modern
hives are provided in three free Advisory Leaflets of the

British Ministry of Agriculture, Fisheries and Food, Nos. 367, 468, 549.

It would be nice to report that application of the known principles of hive construction had led to the universal adoption of a perfect hive. Nothing so simple has happened. In England the small frame was standardised in 1884 and has coloured our beekeeping ever since. In America, no standard has ever been set officially but, by common consent, virtually only one hive is now used, that known as the Langstroth. So too in Australasia and many other countries. This means that in the English-speaking countries, two completely incompatible types of hive have survived: those with lower bee-space in Britain and those with upper bee-space in the USA and elsewhere.

Further division of opinion occurred in the manner of hanging and spacing the frames. The USA adopted from the start a 16-mm lug in a plain rebated box. Studs on the side-bars or wide bars, cut away so as to leave self-spacing wings on the upper 80 to 100 mm, provided the side spacing. In spite of much prejudice, these 'Hoffman' frames are deservedly gaining ground in English practice. The traditional British spacer is a box-type folded tin 'metal end'. In order to take the metal end the lug had to be lengthened to 38 mm and the hives complicated beyond reason to accommodate the long lug, by adding four extra pieces of wood. When a component is placed above these long lugs, which lie flush with the top, it must necessarily touch and be propolised to the lower frames. Each time the hive is opened some of the frames have to be prised away. This upsets both the bees and the beekeeper. In my opinion, gained over many years of working with both types of hive, the upper bee-space system with short-lugged, Hoffman-spaced frames keeps beekeeping pleasant and easy. The others lead to much tedious, unprofitable work, unnecessary stings and neglected hives.

Unfortunately, that is not the end of the argument. Some people like to keep their bees in small brood boxes.

Top bee-space hives – frames clear of upper components

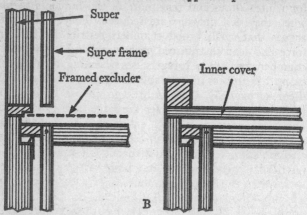

Lower bee-space hives – upper components propolised to the frames

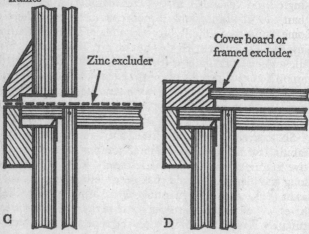

FIG. 6 Adhesions between hive components

They point out that the bees do not need so much room for winter stores and that more of the honey is surplus when the bees' quarters are reduced. The other faction argues that small brood chambers restrict the foraging force and lead to enforced early swarming. Large brood chambers mean more foragers, less swarming, bigger honey crops. I keep bees in both sizes of hive and, whatever the theorists may say, there is little to choose between the takes from either. If I prefer the large hive to the small one it is because it is much easier to work and manage, cheaper in initial cost and upkeep, and requires much less autumn sugar feeding. I find that it is simple enough to restrict the space available in a large hive, but impossible to increase the space in a small hive without doubling up the brood chamber. This means double cost and double work. Further, since one has to frequently move hives to the crops, I know from experience that the largest single brood chamber hive is considerably easier to shift than the smallest double brood chamber unit. There is some claim that double brood chambers simplify routine inspections. I have tried this but do not find it much, if at all, quicker, and it is certainly less reliable than single brood box routines. An efficient and safe doubling method is described in Chapter 7, but the very fact that one has to increase the size of the brood chamber is an admission that it was too small to begin with.

Six currently available hives are presented below in tabular form for the purpose of comparing their respective features. The familiar telescopic 'WBC' hive, for so long popular in England, has been deliberately omitted, as have also the similar Scottish 'Glen' and the Irish 'CDB' hives, all of which share the same design defects. Unfortunately the WBC is foisted on the tyro only too often as part of a 'beginner's outfit'. No beginner should accept one as a gift. Such hives require great experience and skill to manage at all. Most of those one sees around the country have blunted the enthusiasm of their owners and the bees are no longer managed at all. The bees are usually blamed instead of the faulty equipment. All the hives in the table are, as now modified, sound, manageable hives.

Lower bee-space hives

Hive	No. of frames	Frame size mm	Frame spacing mm	Lug length mm	Number of cells in brood box
National:					
Brood	11	356 × 216	37	38	58 000
Super	10	356 × 140	42	38	36 000
Modified Commercial:					
Brood	11	406 × 254	37	16	75 000
Super	10	406 × 152	42	16	

Top bee-space hives

Hive	No. of frames	Frame size mm	Frame spacing mm	Lug length mm	Number of cells in brood box
Smith:					
Brood	11	356 × 216	37	18	58 000
Super	10	356 × 140	42	18	36 000
Langstroth:					
Brood	10	448 × 232	35	16	68 000
Super	10	448 × 140	35	16	
Jumbo:					
Brood	10	448 × 286	35	16	85 000
Super	10	448 × 140	35	16	
Modified Dadant:					
Brood	11	448 × 286	37	16	93 000
Super	10	448 × 159	42	16	

The following further comments on the characteristics of each hive may help the reader to choose.

The National. The most widely used simple hive in

Britain and the most readily available, new or second-hand; its frames fit all usual extractors, but it normally requires two brood chambers. Older versions give trouble with propolis, but recent designs have overcome this to some extent. It is in effect the British Standard hive.

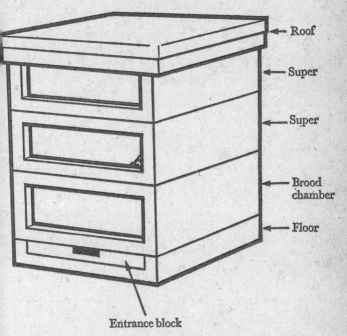

FIG. 7   The BSS National hive

The Commercial. Also called 'Modified Commercial' or simply the '16 × 10', this is a very good hive, except for the lower bee-space. It has the same outside dimensions as the National, and all components are interchangeable with it, except the frames. This hive deserves to be far more popular than it is.

The Smith. This is the only really sensible hive currently available to utilise a British Standard size comb. It is built on the same principles as American hives. The lugs have been shortened and Hoffman frames are standard equipment. The brood chamber capacity is exactly the same as the National and it therefore normally needs two brood chambers. Its top bee-space helps it score heavily over the National.

The Langstroth. The most widely used hive in the world. In spite of its awkward, long brood frame, it deserves its wide popularity. It can just about be used as a single brood chamber hive but, with prolific bees, requires two. In spite of its name it is not the hive designed in 1851 by Langstroth but a very simplified, streamlined version of it. Its one great drawback, as for all American hives used in Britain, is that the frames do not fit extractors designed for the small British Standard frames. Langstroth extractors are now quite readily obtainable but are seldom found on the second-hand market.

The Jumbo. A deeper brood box version of the Langstroth, made so as to avoid the use of double brood chambers. Fully interchangeable, except for frames, with all Langstroth equipment. The frames are the same size as the large Modified Dadant frames, but spaced at 35 mm instead of 37 mm. Ten frames can be used in a Langstroth brood body 50 mm deeper than the standard, thus giving increased capacity for prolific bees. It is a hive which has been regrettably neglected of late, but is enjoying a comeback as the 'deep brood' box of Capt. Tredwell's New Standard Hive system. The Jumbo is my own personal choice of best hive: easy to manage and to transport, and flexible in use.

Modified Dadant or M-D. The largest hive in common use, although its popularity is waning somewhat. For one-man manual operation it is a bit cumbersome, but this hive is widely used by commercial beekeepers. In Central and Eastern Europe a similar Dadant–Blatt hive is popular. It is interesting to note that its brood capacity is slightly smaller than that of a British National with an additional shallow super, which is advocated by many who proclaim that the M-D is much too big!

These last three hives are easier to come by second-hand in the USA and Commonwealth countries than in England.

Assembling hives

When purchasing new hives, considerable saving in cost can be obtained by buying 'in the flat' and assembling the hives at home. Correct fitting of the well-machined parts is simplicity itself. 50-mm rust-proof or galvanised thin nails and some waterproof glue, such as Cascamite, will be required. Assemble the boxes upside down on a clean flat surface to ensure that the four sides are level on top. Nail the corners from both sides, then square up the box and nail two strips of lath diagonally across the base to hold it until dry. If you have cramps, these should by all means be used.

Purchasing frames

Dealers' catalogues list a great profusion of frames to fit each of the hives mentioned. Many of the cheap frames save timber, but waste one's temper. In fairness, it must be stated that the manufacturers do produce a very well made, accurate article. The name of the hive will tell the supplier the length of the frame and top bar; the words 'deep' or 'shallow' will indicate the depth of frame. From then on, the purchaser should stipulate the following:

Top bar: 27 × 19 mm material, wedge type.
Side bars: 9 mm thick Hoffman 35 mm self-spacing (or 27 mm × 9 mm used with converter clips as described below.)
Bottom bars: 27 × 8 mm either single piece or slotted two piece.

Super frames

Individual frames from the supers are not usually handled from the time they are placed on the hive until they are ready for uncapping after the crop has been removed. They need not be made to move so easily and can afford to be spaced further apart, because the cell depth is not governed by the length of a pupal body, as in the brood nest. Simple close-ended frames of this type are known as 'Manley' super frames. The side bars sold in the trade are plain 42-mm wide material. But spacing is only part of the story: the top and bottom bars must be of the same width, 27 mm, so as to act as knife guides when uncapping. Ten of these frames fit all the hives mentioned, except the Langstroth and Jumbo. This probably explains why these excellent super frames have not become popular in the countries where that hive is standard.

Foundation

Worker base, or drone base foundation, can be obtained cut to size for any frame, either plain or ready-wired. Drone base was once popular for supers, but beginners would do well to regard it with the utmost suspicion, except for the sole purpose of breeding drones in a queen-mating apiary—hardly beginners' work.

Three grades or weights of foundation are available: standard, thin and extra thin, which is used unwired for comb honey, where a thin midrib is essential. The standard grade of wired foundation is generally preferred for the brood chambers, while the medium provides a few sheets extra per kilogram for the supers. In the USA, standard brood comb sheets can be purchased with a 10-mm wide tin frame which clinches the wires and allows for great speed in handling. Foundation available in Britain is wired either vertically with crimped wires or in a vertical zig-zag with loops above and below the sheet; a few centimetres of loose wire are left at the bottom of the sheet at either side.

Correct assembly of frames and foundation is necessary if the wax sheet is to remain flat inside the frame. Thin sheets of wax cannot be handled in temperatures below 20°C, and both wax and work room must reach that minimum temperature before any attempt is made to fit the foundation into the frames.

The cost of foundation has driven some beekeepers to revert to the ancient practice of casting their own wax in metal or plastic moulds. The foundations so produced prove to be most acceptable to the bees, but require separate wiring into the frames. In many areas such moulds are available for use at an educational centre. The local Beekeepers Association will always supply information about the facilities in the district.

Assembling frames

The frame components, machined to a good push-in fit, can easily be assembled wrongly. Before nailing, watch the following points:

1 Remove the wedge under the top bar and clean out the rebate.
2 Grooves in the side bars face each other inside the frame.
3 Hoffman side bars always have the V edge on the left facing the operator, when the top bar is uppermost.
4 Nail the side bars from the side into the top bar.
5 Nail in the first stick of the two-piece bottom bar, then fit the foundation and finally put in the second stick. One-piece bottom bars have to be left until last.
6 The hooks on crimped foundation are laid in the top bar rebate so that they hang on the wedge. The upper loops on the zig-zag type have to be bent over at right

angles to bed into the rebate, and the loose wires underneath can be wound round the pins fixing the bottom

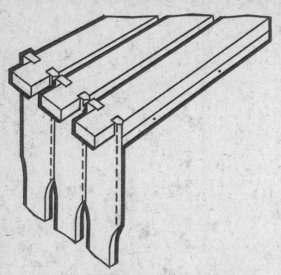

FIG. 8 Correct alignment of Hoffman frames

bar. After nailing the wedge back into its rebate, the job is complete.

Morris converter clips

Plastic clips to fit over the side bars of 22-mm and 27-mm frames and convert them to Hoffman type spacing have recently appeared on the market (Fig. 9). They are a great advance on metal ends, but require care in fitting. When used with new frames, they should be pinned inside rather than outside, where they would tend to upset the bee-space. Their upper edge has to be pushed up level with the underside of the top bar. In this way they will always line up correctly.

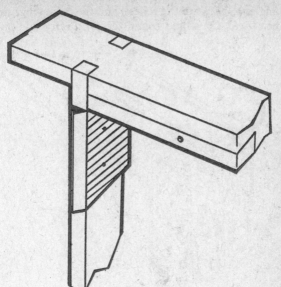

FIG. 9 'Morris' converter clip

There are now a large number of fittings, colour coded for easy reference. The clips made to fit the 22-mm wide frames are, in my opinion, not worth buying, since the frames themselves are intrinsically inadequate. Far better to scrap such poor furniture and replace it with the correct article. When reordering, it is well worth calculating whether the more expensive Hoffman frames are not in the long run cheaper than 27-mm frames, to which must be added the cost of two clips.

Sections

Part of the pleasure of keeping one's own bees derives from producing comb honey in sections for home con-

sumption. Ordinary supers fitted with hanging frames
are far superior to the old fashioned 'section crate' and
have the added attraction that they can be used with
normal super frames if required. Metal dividers separate
the frames and serve the purpose of keeping the comb
cappings a bee-space away from the divider and below the
level of the wooden box. The sections are always sold 'in

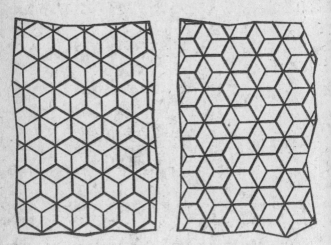

FIG. 10   Correct and incorrect ways for section foundation

the flat'. The hinges or folds need moistening for a few
minutes before folding; one half of the split side can then
be slotted into position. The extra-thin foundation sheets
sold for sections are not quite square and only fit properly
when the embossed cell walls have two opposite sides
vertical, as shown in Fig. 10. When the wax is down and in
the grooves, the top can be stroked over the split top and
the remaining half of the box bedded down. Sections
should always be lined up in their frames with the split
side uppermost.

Bee-ways

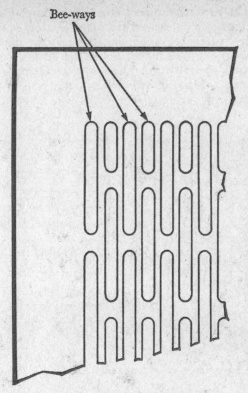

FIG. 11   Zinc queen excluder

Queen excluders

The standard queen excluder in most countries consists of a plain sheet of zinc or plastic, slotted to allow the workers to pass out of the brood chamber into the supers but to exclude the queen and drones with their bulkier thorax. Ventilation is upset to some considerable degree when these excluders are laid flat on top of the combs in bottom bee-space hives such as the National or Commercial. The bees propolise every slot down to the top bars wherever they touch. On the other hand, with top bee-space hives, the excluders sag in the middle and give rise to burr

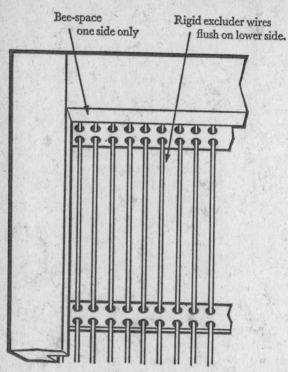

Bee-space
one side only

Rigid excluder wires
flush on lower side.

FIG. 12 Herzog framed excluder

and brace comb above the frames, in addition to their getting propolised to the bottom bar of the frames above.

Clearly, an ideal excluder should consist of a rigid grille with a full bee-space on one side only. So far, I have been able to trace only one which fits the specification exactly and that is the excellent Herzog excluder imported from Germany and distributed in Britain by E. H.

Thorne of Wragby. Slotted zinc excluders can be framed and kept fairly rigid, but they then tend to distort and cannot be flattened out under pressure like the plain metal sheet. The American wood-and-wire excluders are far from ideal but work well in practice. They get so gummed to the supers that they can be lifted clear of the brood box with the supers and only removed from them at the honey harvest.

A wire type of excluder, called the 'Waldron' and made in England, has all the makings of an excellent tool. Unfortunately, the manufacturers have overlooked the basic principle of bee-space and have given their grille half a bee-space on either side. I have not come across any beekeeper who has used them and not increased his vocabulary by several expletives. They can, if met with cheaply second-hand, be reframed to give a clear bee-space on one side and be nearly flush on the other.

Dummy frames

A piece of hardboard, plywood or thin wood cut to the same dimensions as the normal frame, but not more than 10 mm thick, and fitted with a top bar of the usual length and thickness becomes a dummy frame. It hangs in the hive in the usual way and can be used for closing up a gap at the side of the brood chamber. It can form a false wall, when the full complement of combs has not been inserted, and so prevent the construction of 'wild' comb in temporary spaces.

Nucleus hives

Rather than use an expensive full-size hive for housing small units of bees, such as nuclei (see Chapter 5) or small swarms, many beekeepers use small cheaply made boxes to hold three to five brood frames. Various patterns are in use, but they should all have some provision for feeding the occupants, and if they are to be carried around in the boot of a car they need an alternative gauze cover (Fig. 27). Two identical nucleus hives can be placed together under one standard hive roof.

In surveying hives, I may appear to have come down heavily in favour of American as against British designs. This I have done with some reluctance. The truth is that in the USA beekeeping is big business and, as a result, unnecessary and fussy detail and bad design has had to be ruthlessly swept aside. Here in Britain the craft has tended to remain in the hands of worthy amateurs, who keep bees as an attractive and somewhat costly hobby. But if we wish to enjoy beekeeping to the full, there is no reason why we should not use the best professional equipment available.

It is ironical that many sound English beekeepers in the past have done much to pioneer sensible appliances, and at times they have been well supported by the manufacturers. Unfortunately, our traditional reluctance to make any changes at all, even for the better, has put paid to several excellent inventions, which we must now borrow or import from abroad.

However, hives and equipment do not gather honey. That is the province of the inmates of the hives and, in spite of their 50,000 or more stings, it is the bees we must now learn to manage.

4

Handling Bees

If you watch an experienced beekeeper open his hives, examine the combs, shake off the bees and steadily work his way through the apiary, you might be forgiven for thinking that bees were the most harmless creatures in the world. This degree of skill in handling can be achieved by those who are willing to take the time to learn and practise the techniques.

At the outset, it must be admitted that no book can replace practical tuition under a competent beekeeper. I cannot give better advice than to recommend joining the practical beekeeping classes provided by local adult education services. Where such facilities are not available, an adequate substitute can be found in the local amateur beekeepers' associations. Most of them include a number of very good beekeepers who will help, teach and give confidence, particularly if the student makes a point of requesting this kind of tutelage. Even if the advice should conflict in some details with that given here, it would not necessarily mean that either was wrong but merely reflect a different approach as between two different practitioners. Beekeeping is not a craft to be rushed at. Time taken over the early stages will be amply repaid in ensuing years.

Special gear is required for handling bees, and it is pointless to buy equipment which does not perform as well as it should, or fails at crucial moments. We must take into consideration the basic fact that bees sting to defend their colony against attack or interference, and our efforts at management constitute an interference which they resent.

The veil

Always use a veil. Later on you may be able to dispense with it on some occasions, with some bees. But confidence can only be gained and kept when one's face, eyes, nose, mouth and ears are safe beyond doubt. A veil should be light, quite bee-proof and kept well away from the face. The lower edge of the net should be worn running up and over the tops of the shoulders, not pulled down over the arms. Should a bee manage to creep into a veil, there is no need for alarm: the bee's frantic buzzing shows that it is terrified and wants to get out. If it is found to be too disturbing, it can be crushed inside the veil at an opportune moment. A number of different veils, or combined hat and veil outfits, are now on the market to suit every taste, but for the small beekeeper the heavy metal veils are cumbersome to put on and tiring to use.

Bee suits

Of recent years bee suits have become fashionable and, certainly, the right kind can be quite effective in freeing one's mind of any doubts about the suitability of the clothes worn when attending to the bees. A white denim boiler suit, preferably with a zip front and zipped or sewn-up pockets, presents an impassable barrier to marauding bees. Their only access would be up the trouser legs or sleeve. If desired, the trousers can be tucked into a pair of wellingtons and the cuffs closed with a pair of elasticated false sleeves from the wrist to halfway up the forearm. Now, only the hands are bare.

Gloves

Gauntlet gloves are worn by many timid beekeepers and by most professionals. This apparent paradox can be easily explained. Some self-professed beekeepers cannot overcome their fear of bees; they have never really learned beekeeping skills and their first blundering efforts with gloves have not taught them the ease with which operations can be carried on with bare hands. The commercial beekeeper, who may have to open hundreds of hives a week, is protecting his hands from the clinging, sticky propolis more than from the stings he is accustomed to. But he does know how to handle bees, and is most unlikely to have learned his beekeeping with gloves on. Handling bees demands a steadiness and accuracy in procedure that is completely jeopardised by all protective gloves. After bees have been successfully managed with bare hands for a year or two, gloves can be resorted to, if it is still felt that they can be of any use, or for special occasions.

The smoker

So far, we have considered the equipment necessary to protect against stings; it is important now to consider how to persuade the bees not to sting at all, and the simplest and most efficient tool for this purpose is the smoker. Of the various types available, the only one worth buying is known as the bent-nose smoker. The firebox can be made of steel, which corrodes, or of copper, which does not. Fuel for the smoker should be some slow, cool burning material—old sacking, rotten wood, peat or corrugated cardboard. Combustion depends on the material being kept dry and, to maintain a steady draught, the holes in the 'fire grate' and nozzle have to be scraped clear of coke from time to time.

The smoker can be lit, even in a light breeze, if a twist

of lighted paper is held in the fire-box with a piece of fuel held over it until it begins to burn. The bellows are then worked until the material is well alight and can be pushed into the firebox, lightly enough for a draught still to blow

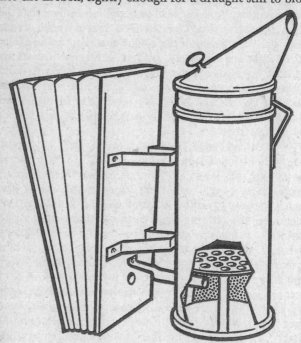

FIG. 13   The bent nose smoker

through when puffing. Work the bellows until a steady stream of smoke can be relied on. When the smoker is stood upright, it will smoulder away steadily until required, but if it is laid on its side, it will go out after some minutes. A cork or a plug of green grass stuffed into the nozzle will make it go out even quicker, when it is no longer required. I make it a rule never to empty my smoker while it is still hot, but leave the emptying and raking out until it has to be used next time. Smouldering fuel, left to blow about in the open air, can be a great fire risk.

Disregard anyone who tries to convince you that smoke is not necessary, or is harmful to bees. The truth is that the alternatives are always worse. Too much smoke upsets bees; there can be no doubt about that. But no smoke at all, or chemical subduers, lead to concerted attacks by the bees. Every one that stings dies, which is hardly kind to them. On the other hand, moderate smoking subdues the bees, holds their aggressive tendencies in check and allows them to resume their duties within minutes of closing up the hive.

Why smoke should affect bees in this way is still unclear, but in practice it is found to work. Within one or two minutes of directing a few puffs of smoke into a hive, a considerable number of bees gorge themselves on the open honey-cells and lose most of their aggression. It has often been asserted, probably correctly, that gorging is an instinctive reaction of self-preservation for the bee colony against forest fires. According to this theory, as many bees as possible would load up with honey and decamp to a new home, as soon as the smoke became too intense. I recently had the dismal opportunity of visiting the scene of a heath fire, in which four out of a group of sixteen hives were completely burnt out, and four others severely charred by the blazing heather and bracken, yet not one of those colonies had decamped. Time has perhaps attenuated the instinct, or some other explanation should be sought. Bees do, however, show very reduced aggression after smoking and always run away from a direct draught of smoke. By taking advantage of this reaction, we can keep them under effective control during manipulations.

The hive tool

Bees glue down with propolis any contacting surfaces in

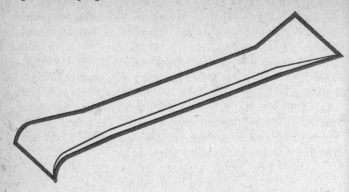

FIG. 14   The hive tool

the hive structure, and whenever we wish to open it up, some degree of force is required. A screwdriver or blunt chisel can provide the leverage, but both these implements tend to damage the soft wood of the hive boxes. A hive tool does the job far more efficiently and can be useful in the house in many ways; I often wonder how non-

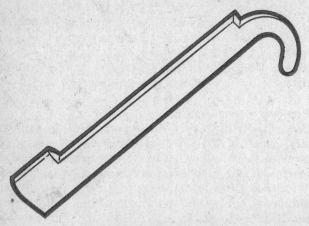

FIG. 15   The h-j hive tool

beekeepers ever manage without one. The usual type consists of a broad steel blade, thinned at one end and widened at the other, which is turned over about 12 mm to form a scraper. Another design which is very useful in hives fitted with the recommended Hoffman frames is now available. It is called the h-j from its appearance. The main lever is relatively narrow, but broadened out near the end on one side to form a blade; the other end is turned over to form a hook, specially designed to lever out the first frame from a tightly packed hive.

Stings

However valuable these physical methods of protection may be, by far the most important item in our armoury is a good understanding of the factors which cause bees to sting. Unless we acquaint ourselves with these at the outset, we are bound to provoke the bees and suffer the consequences. Bees die after they have stung, and so do not use their ultimate deterrent lightly but only in response to a stimulus which provokes the stinging reaction. Unfortunately, these stimuli tend to be rather numerous, and it is worth while spending some time examining them in detail. The bees' reactions are always defensive and if the word 'aggressive' is used, it must be understood from the subjective viewpoint of the beekeeper.

Bumping or jarring the hive

Bees hear low frequency vibrations transmitted through solid materials. If we wish to approach bees quietly, remember that we can talk, shout or sing to our heart's content; the bees cannot hear it, and for them we are as silent as the grave. If we then approach the hive and merely put our smoker or hive tool on the roof, we have made a noise which alerts the guards, and by the time we have started to open up, they will be waiting to attack. Any kind of knock, bump or tapping on the hive should be avoided prior to opening up.

Stings lead to more stings

When a sting is torn away from a bee, a pheromone is released which stimulates other bees to sting: it acts as an alarm signal. It is not necessary for the sting to be felt by the beekeeper; it may well be in his clothes and unsuspected, but for a considerable time afterwards the scent persists and continues to provoke other bees. Bee suits need to be kept clean for this reason alone, and the difficulty of washing leather gloves, which retain old stings, is another reason for not using them. If the hands are stung, the sting should be scraped out with the hive tool or a finger nail and the area smoked to hide the alarm scent.

Bee blood

Bee blood also releases a pheromone whenever a bee is crushed. Naturally one does not like to kill these admirable creatures, and the utmost care needs to be taken in handling frames heavy with brood and stores. A bee's skeleton is tough, but not strong enough to resist the pressure of several pounds weight between two pieces of wood.

Sudden and jerky movements

Any kind of jerky or rapid movement crossing a bee's field of vision provokes a stinging response, and this obliges us to learn how to handle them in a calm, deliberate manner least likely to upset them. Other implications arise, too. When a bee, through some fault of ours, does happen to sting, our natural reaction forces us to jerk away the wounded hand, or to smack vigorously any other wounded part, with the inevitable consequence that the movement incites other bees to attack. The tendency to react in this way must be overcome at all costs. A little stoicism with one sting is better than a lot of self-expression with many. Although the beekeeper is never destined to succeed entirely in his efforts to avoid being stung, he can, with self-control, considerably reduce the number of his painful failures. A proper reaction to stinging consists in 'freezing' for a moment or two and then very slowly indeed, reaching for the smoker and blowing a few puffs around to disguise the alarm scent, while the sting is removed. It is also counter-productive to try to beat off a bee which approaches near one's face by flailing around with the arms. It may only have been looking for a temporary resting site, but is made to realise that it is in the presence of a determined enemy and responds accordingly.

Robbing

When hives are kept open for too long a period, bees from other hives in the apiary, attracted by the smell of honey and the large undefended opening, may well start robbing 'over the shoulder' of the operator. They alight on the alien combs to steal, and the legitimate owners, not unnaturally, become aggressive and attack all and sundry. This happens more frequently than many people think, and is especially prevalent late in the season. It first comes to notice when an apparently calm and well-controlled colony under examination suddenly rises in anger for no apparent reason. The reason is there, however, and is due to the beekeeper taking too long over the job. His only remedy lies in closing up as quickly as possible and restricting the hive entrance for a while, to give the guards a chance to restore peace and quiet.

Colours, smells, static electricity

I have grouped these together because, in practice, the stimuli involved all derive from clothing or the operator's person. Woollen materials, tweeds, sweaters and knitwear in general attract stings invariably. So do dark blue or black clothes—and do not forget that red is black to bees. I sometimes think that one is better protected with nothing but a loin cloth than with a tweed jacket, but I must admit that there are better alternatives. Animal fibres appear to diffuse a scent, discernible to bees, which arouses their fury. Their reaction to nylon is less clear, but in the absence of experimental proof I presume it must be caused by the static electricity engendered by this material through friction during quite normal movements. Breath occasionally upsets them, but I have never found it necessary to hold my breath during a lengthy inspection. It is sufficient to hold the comb high and direct one's breath below the frame, or to keep one's head about a foot away from the hive. Some perfumes have the reputation of producing similar effects, but while working in the apiary with ladies I have never noticed that the most seductive of scents had the slightest effect on the bees. Some say that human sweat makes them sting, but I have found this to be totally untrue, and I can speak from experience.

Effect of stings

The sting poison of bees is a complex of enzymes and proteins, so that the traditional blue-bag cure has no effect whatever. Although a minute amount is injected, its effect on some people is out of all proportion to the quantity carried in the bloodstream. Most of us quickly develop a form of immunity from the painful after-effects and look upon stings as a temporary annoyance, quickly alleviated by the removal of the actual stinging apparatus left behind in the skin. Some take longer than others to achieve tolerance to stings, and for them relief can be obtained by early application of an anti-histamine ointment or spray, both readily available from chemists' shops.

However, there are just a few people in whom the sting poison produces a quite opposite effect, by building up a serious allergy with possibly fatal consequences. For them the pleasures of beekeeping must be confined to learning from books or the media. Should the first contact with bees lead to dizziness, trembling or temporary unconsciousness, immediate medical advice should be sought and all further physical contact with bees avoided. The proportion of people allergic to stings is very small indeed, somewhere in the region of one in half a million, but the fact remains that an allergy can exist, or even arise after years of apparent immunity. For this reason it would be wise not to start beekeeping and buying bees until a season's work in the apiary, under some supervision, has shown that physically and temperamentally the bees and the aspirant will get on together.

It is to be hoped that information about stings has not led the reader to suppose that beekeeping is all pain; that would be most untrue. Few hobbies or crafts can provide such deep satisfaction and sustained interest over many years, nor such pleasant outdoor relaxation. Learning to manage such fascinating creatures induces self-control, broadens the mind and improves the body, and knowing about stings is the first step towards avoiding them.

Opening the hive

Make sure that your smoker is well alight and can provide a steady stream of smoke at a moment's notice. Start working from the back of the hive if conditions allow, or at one side if that should be necessary, but avoid standing in front of the entrance where you would be in the flight-line of returning foragers. Hold the smoker in your left hand and your hive tool in your right; unless, of course, you happen to be left-handed. Remove the roof gently so that its sides do not bump the hive, and place it upside down behind you. Many operators place it in front, but this encourages returning pollen gatherers to enter the supers, which will soon be placed over it, and to unload their pollen there, above the queen excluder.

Then force the hive tool about an inch between the excluder and the super, press down on the lever to prise the first gap and immediately puff smoke into it. Lower the box gently and repeat the operation on the other side, including the puff of smoke. The propolis should now be broken sufficiently to enable the super to be twisted round to uncover all four corners. Give each one in turn a puff of smoke, and lift off the super or supers complete with the crown board and place them diagonally on the roof at the back. Then prise off the excluder and shake it over the hive to dislodge the bees; place it against a side of the hive away from the front entrance, or flat on the cover board. As soon as the brood chamber has been uncovered, one or two puffs of smoke can be directed across the frames to drive the bees down.

Smoking

It is to be noted here that I place the emphasis on subduing the bees immediately the hive is opened and maintaining them under control during the whole operation. The quantity of smoke used is, and must be, relatively small. The timing of the puffs matters far more than their volume: a puff in time saves nine. The effect of smoke can be seen and heard. Too little fails to produce any result, but the right amount is followed by an audible, temporary buzz and the disappearance of the bees into the depths of the hive. Sometimes during the manipulation bees start welling up at the corners and over the combs; a gentle reminder with the smoker ensures that the beekeeper, and not the bees, remains master of the situation. Once control has been lost, it is difficult to regain with hundreds of bees milling round in the air. Bees can be controlled in the hive, but not out of it. Too much smoke produces an angry roar of protest and sometimes even a mass exodus. The right amount of smoke is produced when a single, definite draught is directed firmly to the exact spot or spots where trouble is developing and the bees respond with a definite, short buzz.

Removing the first frame

With the bees well under control, the hive tool can be inserted between the hive wall and the nearest side-bar. Lever the frames inward, hard enough to move the whole brood nest over into contact with the opposite wall, and repeat on the other side of the frames to straighten the block. This operation leaves the maximum possible space for removing the first frame, without damage to the comb or to bees. Now insert the bent end of the hive tool between the last and last-but-one frames and twist it to break the propolis without jerking. The smoker should be placed to hand upright on the ground and the hive tool held between the third and fourth fingers and the palm in such a way as to leave the thumb, first and second (middle) fingers free. Grip the lugs firmly between the thumb and first finger and gently raise the whole frame, keeping it central and level in the space available. As soon as the frame is clear of the hive, slip the middle finger under the lugs and bring the balls of the thumbs down onto the side-bars to steady the hold and improve the grip. In this position the frame can be examined on one side, but it must be held over the hive at all times whilst bees are on it.

Turning frames

Combs, heavy with stores or brood, cannot be held in a horizontal position without serious danger of collapse, especially in warm weather when the wax is soft. This means that we have to use a technique for turning them round that maintains them, throughout, in the vertical position. Fig. 16 shows the sequence to master before we actually tackle a full frame with bees and an open hive. An empty frame can serve quite well to practise on. First transfer the weight of the comb back onto the finger tips, and bring the right hand directly below the left; the bottom-bar will then be on your left. The movement is from the elbows and not the wrists. Gently swing the frame

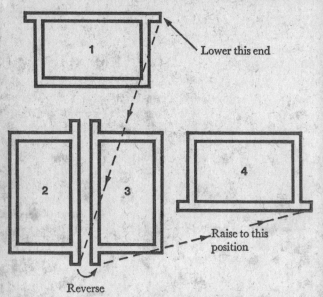

FIG. 16   Turning the brood comb

round away from you, so as to bring the bottom-bar to the right, while still maintaining a good grip. Bring the left hand down, and slide the thumb and first fingers up the side-bar to steady the frame while you examine the other side (Fig. 16).

An exact reversal of this routine brings the frame back to its original position, ready for replacing in the hive. The first frame removed can, with advantage, be shaken clear of bees over the hive and leaned, end up, near the front, thus providing more space for freeing and removing subsequent frames. If, however, it contains brood or eggs, do not shake it, but lay it gently over the tops of the frames away from the side where you are working. Care must be taken to avoid 'rolling' bees between adjacent combs during the lifting process; they quite rightly object to it and there is always grave danger of damaging a valuable queen.

Shaking bees

Reference has already been made to shaking bees off a comb, and this manipulation requires a little practice before trying it out in real earnest. Take the whole weight of the comb on the middle fingers under each lug, and hold the frame a few centimetres above the hive; then strike the top-bars with the balls of the thumbs, so as to jerk the comb sharply downwards and dislodge the bees. More than one jerk may be needed. Honey frames, excluders and cover boards can be shaken by gripping them firmly in one hand and thumping them firmly with the other fist. This method proves very effective in removing bees but is too violent for brood frames, since it can also dislodge eggs and larvae.

Manipulating cloths

During the course of these operations, with light streaming in from above, the bees may attempt to rise and repel the invader. An occasional puff of smoke soon restores calm, if done in good time, but much smoking can be avoided if two manipulating cloths are used to keep the frames covered. They are simply two pieces of twill or sail cloth about 55 cm × 55 cm with a strip of lath tacked to opposite ends. These can be rolled or unrolled over the brood chamber as the inspection proceeds across the hive, in such a way as to leave a gap 8 to 10 centimetres wide to work in. They have the added advantage, in bright weather, that the movement of the arms across the top of the hive does not excite the bees to sting. They are also very useful in driving bees away from the hive top prior to closing up, or as temporary covers if an open box has to be left for some minutes. Some beekeepers use cloths soaked in a carbolic solution, but I have never found the carbolic to be beneficial; on the contrary, it is even more disruptive to colony routine than smoke and takes longer to get over, because the smell absorbed by the wax remains in the hive after it has been closed up. Also, late in the season, carbolic incites bees to attack instead of subduing them.

Finding eggs and queen cells

It is seldom necessary to open and disturb a hive except during the swarming season, and the whole purpose of examining combs then is to ascertain the presence or absence of eggs and queen cells. We shall deal with swarm control in a later chapter, but the beekeeper must start learning to recognise eggs as soon as he or she first looks into a hive. If glasses are normally worn, they should be used for inspecting combs. Eggs can only be found in open cells, but the sealed brood cells can be the starting point for our search. The nest expands in concentric circles in the spring and at this time of year the oldest bees will be in the middle of the brood patch and the youngest near the perimeter. A ring of eggs fills the outermost circle of the nest. Several combs may have to be examined because the central combs may contain no eggs at all. To see eggs it is essential to look straight down into the cells. If the eggs are present, they can be seen as small whitish streaks standing out from the cell base. Care has to be taken not to confuse eggs with light reflections from the cell bottom. Familiarity with their three-dimensional appearance will soon enable the beginner to recognise eggs

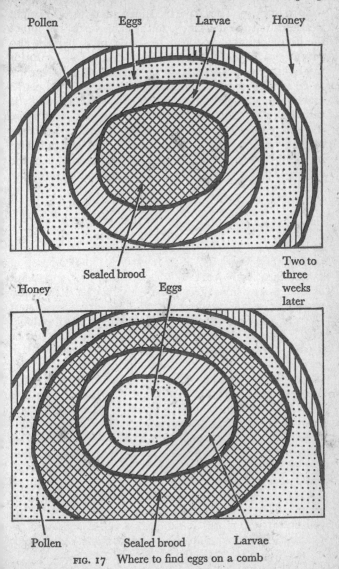

FIG. 17 Where to find eggs on a comb

with certainty, and when he does see them he knows that there is a laying queen in the hive.

Queen cells are easier to recognise, but we must be sure of the meaning of the words 'queen cell'. Quite early in spring, bees build a number of vertical cup-like cells in

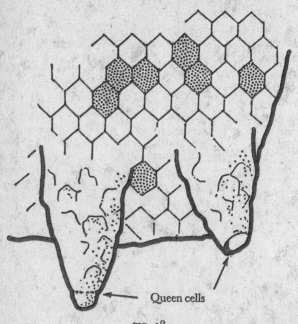

Queen cells

FIG. 18

various places throughout the brood nest. These cups may never be used and they are not what we mean by a queen cell, since they contain neither larva nor royal jelly. We can only be in the presence of a queen cell when the cup contains a larva floating on royal jelly and the bees have begun to work on the edges of the cup, or have already considerably lengthened it. Examining for queen cells, therefore, entails looking into the cups and checking to see whether they are occupied or not. When it is necessary to find all the queen cells in a hive, the bees must be shaken from every single comb, and its surfaces, edges and hollows thoroughly scrutinised.

Finding the queen

For some essential operations, the queen has to be found, and the task of sorting out just one bee among so many presents the beekeeper with a challenge which must be met. First, ask an experienced beekeeper to show you a queen on the comb and to help you become familiar with her appearance and behaviour; her behaviour, at least as much as her outward appearance, will help you to distinguish her from the workers. Avoid using too much smoke when opening the hive, as this will drive her into some remote corner. The best time of year for finding queens, if you can afford to pick your time, is in spring, when the brood nest is expanding rapidly; there are then less bees on the great slabs of brood, and the queen is so busy laying that she is scarcely disturbed by your intrusion. She is seldom to be found on the outside combs which contain no brood, but more likely to be busy laying on either of the last two combs of the expanding brood area.

Many beginners make the mistake of bringing the comb too close to their eyes, so that they can only see part of it at a time. Hold the frame far enough away to see the whole of it in sharp focus, and if the queen is on that comb face she can often be picked out at once. If not, turn the comb round and examine the other side. Always look first on the side of the comb which has just been detached from the nest, because she may have run away from the light when the previous comb was lifted. Before putting the frame back, carefully examine the edges, and the bees on the outside of the frame. If she is not found on any of the frames, go back, retracing your steps across the hive; she will be found eventually. Some queens are notoriously shy and difficult to find, but going back over combs previously examined usually leads to taking her unawares. Some experts have special recipes for finding queens, but I must admit that I can never make them work and prefer to rely on finding my queens at the right time of year and by systematic examination of the combs. With some experience, the queen can be found in a very short time by choosing the most likely combs to inspect first.

Before replacing the comb with the queen, make sure that there is plenty of space on each side and that the queen is not on the wooden frame—she might get crushed between the frame and the hive, and if you have not yet found her every comb must be treated as gently as though she were on it.

Closing the hive

When the examination is over, the hive should be closed with as much care as it was opened. Replace the last comb, and if you have left one leaning against the hive front, put that one back too. All combs must be replaced exactly as they were before opening up. Now lever the whole brood nest over to one side, pressing on the outermost side bars alternately so that the V edges on the frames cut through any propolis. Only in this way do the frames line up at the correct spacing. If this creates a gap on one side lever the nest back en bloc from the other side, so as to leave equal spaces between hive and frames at each side. Drive the bees away from the top with a puff of smoke, or lay the cloth across the top for a second or two, and replace the excluder. Put the supers back slightly askew at first and then twist them back into line; this minimises the area of contact and prevents bees from being crushed. Feel with the fingers at the joint to discover any overlap, which should be corrected. If this is done at diagonally opposite corners, the upper and lower boxes line up square. Put the roof back gently and check that it is seated level on the top of the crown board.

These procedures for opening and closing hives and for inspection of their contents may have seemed distressingly long and complicated as you read them. In fact, a little practice with an empty hive will show that the skills can be learnt quickly. Good habits formed at the start result in smooth, easy working and turn beekeeping into the pleasure that it should be.

5

Apiaries

Apiary sites

Most of us when choosing an apiary site put up with Hobson's choice, but this can lead to an abrupt end of our beekeeping activities immediately the neighbours discover hives at the bottom of the garden. With our ever-expanding towns, the extent of gardens diminishes in proportion to the rise in land values and, with increasing proximity, trouble with timid, non-beekeeping neighbours makes life difficult. And yet those suburban areas can provide considerable crops of honey which it is a great pity to lose. The nuisance to neighbours can be minimised by judicious placing of the hives.

At all costs, avoid setting up the apiary in a frost pocket with no air drainage. Choose a spot in the garden where the cold heavy air can flow away. Failure to check on this essential point leads to bees being kept in what amounts to a sub-arctic climate, retards their development and weakens their resistance to disease. A sheltered situation, where the sun can shine on the hives in the morning and the evening, and with some tree shade during the middle of the day, is the ideal to be looked for, even if some compromise has to be reached in practice.

Which way to face the hives presents less of a problem than some people think. The sun does not have to shine in through the entrance hole—in fact, the construction of the hive effectively prevents it from doing so—but it

Hedge

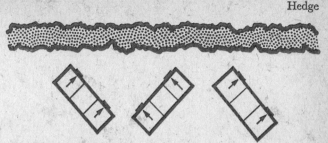

Double hive stands positioned for compactness

Hedge

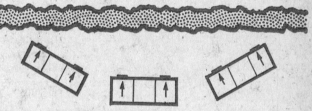

Double stands sited for easy working

Single hives for minimum drifting

FIG. 19   Siting hives (arrow indicates entrance and flight direction)

does have to shine on one of the four sides to warm the hive. This is one of the many reasons why modern single-walled hives are preferred to the old telescopic double-walled varieties. Avoid pointing the entrance into the bitter winter winds; in Britain, this means the quarter of the compass between north and east. If there is any danger of people having to pass or work near the hives, point the entrances towards a hedge or fence, about 2 metres away from it, so that bees coming and going are obliged to rise high enough to clear the heads of bystanders. If only one or two hives are kept, they can be placed side by side at least 30 cm apart, but if more are contemplated, avoid having them in long rows. Four hives can be placed in a shallow V formation, two on each side. If you keep more than four, place them in bays in an open U formation, with two hives at the base and two pairs slightly angled out on either side.

Care in placing the hives will repay the beekeeper in two ways. Bees seem to recognise left from right and seldom enter the wrong hive of a pair, but they are confused by a long row of hives and the prevailing wind tends to drive heavily laden foragers into the hives at the leeward end. They also read the angle of the entrance in relation to polarised light from the sky. If several bays need to be used they will be far enough apart for drifting to be negligible between adjacent groups. The other advantage of the shallow bays is that a clear space is left at the back, well out of the flight-line of bees. Manipulations can be carried out more comfortably and the scent of open hives attracts less attention from robber bees.

Hive stands

Modern hives are not made to stand permanently on the ground. They need to be raised 30 to 40 cm above the level of the grass to prevent the floor from rotting and herbage from blocking the flight hole too frequently. This also eases the beekeeper's task when he bends over the hives. If they are raised too high, however, trouble may be experienced in lifting off heavy top supers in good seasons. I prefer to use stands made to hold two hives at a time. Two 5 cm × 5 cm runners at least 1·2 long can be placed on top of breeze blocks, stood on edge, or more elaborate stands can be made. In woodlands where wood-ants abound, stands have to be made with legs standing in tins, clear of the sides, containing a small quantity of old engine oil or creosote. I have known these pests to rob out a weak stock and take the hive over completely as an adjunct to their ant heap. If the two-hive stand is adopted, place the hives at the extreme ends leaving a space in the middle, and remember that bees at one end hear bumping and jarring transferred from hive to hive via the stand. Be extra careful to avoid alerting the second hive while you are working on the first.

Overstocking

Beware of putting too many hives in any one apiary. Some years ago I kept twenty-four hives in a favourable subur-ban situation and my honey crops, in spite of the forage, were consistently low. I eventually reduced the number of hives to twelve and the crop per hive doubled, giving me the same total return with half the work. I subse-quently found that several apiaries existed in the vicinity and the combined forces of all the hives were competing for the available nectar. The experience taught me to take into account not only my own bees but also the total owned by other beekeepers in the same area. Each loca-tion in an average year can allow just so many hives to be kept economically and no more. What that number is can only be discovered by experience and experiment. With present-day agricultural practice it is seldom pos-sible to keep more than twenty hives on one permanent site, and often no more than twelve.

The flora

Bees gather their honey and pollen from flowers, and it is just as important to study the flora in the half-mile radius round the apiary as it is for a farmer to grow grass and fodder for his cattle and ensure a constant supply of food. Because bees store their food and use it as occasion demands, the beekeeper does not have to fill feeding-troughs every morning and every night, but this does not relieve him of the responsibility for ensuring that the surrounding countryside can provide for the seasonal needs of his livestock.

The development of the annual cycle in a bee colony depends strictly on the annual cycle of the flora in the foraging area. Thus the first big increase in egg-laying will correspond with the availability of early pollen from hazel, snowdrops, crocus or other early sources. Later the willows provide both nectar and pollen, followed by soft fruit, top fruit, holly, maples, sycamore in succession, to provide an early flow. A gap in June, sometimes filled by hawthorn, an erratic and unreliable honey plant, is followed by the main honey flow of clover, field bean, oil seed rape, corn marigold and lime trees. Late flows can come from willow herb and ling, the heather of European moorlands. This sequence applies in the area where I keep my bees, but not all these plants grow in abundance in the same locality. It remains up to every beekeeper to study the flora round each of his apiaries and to modify his management accordingly. In other parts of Britain and in other countries the flora may differ in many ways, but most people have some knowledge of the sequence of local plants and can supplement deficiencies by judicious planting in their own gardens or even by giving cuttings and seeds to their neighbours to extend the acreage of useful flowers.

Stocking the apiary

For your first hive of bees, choose the hive you want to use and avoid the temptation of accepting a free gift of doubtful bees in unwanted, dilapidated equipment. You may, of course, be lucky, but nine times out of ten the problems of setting up neglected or sick bees and transferring them into the hive of your choice deters most experienced beekeepers; for a beginner to start off in this way defies common sense.

Bees can be purchased from many appliance manufacturers, commercial beekeepers and local association members, who are responsible enough to ensure that your first stock is of good quality and headed by a young, vigorous queen. Order your bees well in advance and do not show too much impatience if they do not arrive on the date you specify. They develop in an annual rhythm which can be advanced or retarded by unforeseen climatic variations from year to year. Suppliers usually notify you by 'phone, wire or air-mail a day or two before delivery and leave you to arrange for collection at a predetermined place. Buying bees is one of the many disciplines you have to learn and can only be fully appreciated by those who come eventually to sell them.

Bees can be sold as queens, swarms, packages, nuclei or stocks. Queens cannot be used for stocking a hive, only for re-queening an existing colony. Swarms are bees headed by a queen, usually an old one, without combs; if their source is known and can be relied on, they constitute the cheapest way of stocking one's first hive, but they usually arrive too late in the season to provide a surplus during the first year. To hive a swarm is easy enough and will be dealt with in Chapter 7.

Installing package bees

Package bees, like swarms, are sold without combs; they have a separately caged new queen and worker bees in sufficient quantity to build up a stock rapidly. The package is sold by weight of bees and should contain a minimum of $1\frac{1}{2}$ kilograms. At about 9000 to the kilo this should amount to some 13000 to 14000 bees, which build up to a strong gathering force when they are delivered early enough in the season. A package is introduced in the following simple way. Prepare 10 litres of heavy syrup (see Chapter 9) and sprinkle some of it over the mesh of the travelling cage an hour or so before the transfer, which is best done in the late afternoon on a fine day. Take an empty shallow box and place it over the hive floor, temporarily removing the brood body, which should be filled with frames of drawn comb for preference. Remove the queen cage and tear off the strip of card or paper which covers a plug of candy in one of its ends, then pierce the dry candy with a panel pin and sprinkle a few drops of syrup on the mesh of the queen cage. Remove the small feeder from the package, which you can then open, and shake the bees into the empty shallow. Put the brood chamber above it and over the bees; wedge the prepared queen cage between two top-bars. I like to put a few bees round the cage to 'call' the others up; I usually find enough in the shaken package for this purpose. Replace the cover and immediately place a feeder with the remainder of the syrup over the feed hole. Make sure the entrance is open but in the reduced position and close down the top. Next day the shallow box and the queen cage can be removed. If by any chance the queen has not been released, remove the gauze and the queen will run down and join the bees. Be sure not to stint on the sugar syrup when installing a package. Breeding will start as soon as the bees have had a chance to gather some pollen and not before, but they need syrup to put them in good heart, and may need another 5 litres or more of syrup in a week or two if there is no appreciable flow. Package bees need to be settled two to three months before the flow if they are intended to produce a surplus in the same year.

Installing a nucleus

A nucleus usually consists of three or four combs together with a young queen, her brood of all ages, and a reasonable store of food, both honey and pollen. I would always recommend a beginner to start with a nucleus, because he or she can then be gently eased into beecraft. Experience will grow with the size of the colony and, if sufficient care is taken, a nucleus in early summer should develop into a self-supporting unit before winter and start off the following year as a powerful honey-gathering unit.

When the nucleus arrives, it should be placed in the apiary on the stand where it is going to stay; open the entrance so that the bees can fly and let them settle for an hour or two. When you are ready for the transfer, move the travelling box and bees to one side and put the hive back on its permanent stand, removing from it frames to the same number as those in the nucleus, plus one extra. Puff a little smoke through the gauze onto the bees as the cover is unscrewed or prised from the travelling box. If the queen is caged, set the cage on one side and carefully lift out an outside comb, place it in the space left in the hive, and transfer the rest of the combs one after the other, keeping them in the same relative positions as they were previously. Tip the travelling box upside down over the hive and shake or knock out any remaining bees over the combs. Put back the extra comb to complete the brood chamber; tear off the paper on the end of the queen cage and pierce the candy, then wedge the cage between two top-bars. Put on the cover board and a 10-litre feed of syrup, and close up the hive. Check that the entrance block is open in the restricted position and put back the roof. Two or three days later, remove the queen cage. A nucleus should get off to a better start than a package because it already has brood and pollen, but it will still need two months before the flow to produce any surplus at all the same year.

Installing a stock

A quicker build-up still is obtained by purchasing a stock of bees, and this means anything larger than a nucleus. In practice, stocks consist of six or eight frames of bees and they are transferred in exactly the same way as a nucleus.

In all cases of transferring, the entrance should be left open, but reduced, and a 10-litre feed of syrup given. If the feeder used is smaller, the feeds have to be repeated as soon as the feeder is empty. When bees are sold they are nearly always deprived of two or more full combs of honey to facilitate transport, and this must be replaced to maintain a balanced colony. The smaller nuclei need the extra food to make wax for comb building and developing the nest.

Development of the colony

Once a week check on the developing colony by removing the cover board and glancing at the seams or spaces between the combs. Look for the presence of sealed honey and bees at work. When the stores and occupied seams reach the inner faces of the outside combs, the bees should be opened for examination for swarm control, if that stage is reached within a month after midsummer. There is absolutely no need to pull frames out before then, unless the brood nest has stopped expanding and you want to know why. You can find out by looking at the combs to check whether lack of food has halted growth, in which case feeding will soon put the matter right, or whether there are no eggs; later chapters will help to determine the cause and remedy in this case. A newly installed swarm or package may actually diminish for the first three or four weeks until the first brood starts to hatch out, but it then develops rapidly if all is well. There need be no cause for alarm unless the colony fails to expand after four weeks from transfer.

Out-apiaries

Any apiary situated some distance from home is an out-apiary and it requires special consideration, whatever may have been the reasons for setting it up. If you must keep bees away from home, keep them in a good producing district. I am frequently told by beginners that a particular site must be good because it is in an orchard. But a few neglected apple trees at best contribute to the spring development of a colony, and at worst, if the trees are cared for and regularly sprayed, the insecticide used may deplete or destroy the bees. Sites along river valleys, in upland permanent pastures on chalk or limestone, on the edge of woods or in large parks provide the best sites that can be found under our modern conditions of industrialised agriculture. Arable land put down to cereals or temporary grass leys provide little, if any, honey unless the weeds that manage to survive the chemical weedkillers supply some forage.

Certain dangers attend the placing of bees in out-apiaries, not the least of which are juvenile humans out for fun . . . and mischief. I once had an apiary of sixteen hives not far from a school. One bitter winter when the snow lay on the ground for six weeks or more I did not venture out during that time to look at the hives; when I eventually did I found them all pulled open and lying on their sides and the bees dead, with sixteen lengths of string stretching from the hives to the school fence. It must have been great fun, but it broke my heart. If cattle graze in the vicinity the hives must be fenced off with barbed wire or stock wire, or if situated near woodland the hives may have to be protected with wire-netting cages to prevent woodpeckers from pecking large holes in the hives and eating up the bees inside.

In placing the hives, be sure that you can drive the car or van alongside, on a hard track right up to the apiary, the whole year round. Carrying a full hive or full supers across 100 or so metres in a muddy field is definitely not to be contemplated. Quite frankly, I prefer the bees to have to fly a bit further to the crop than have to cope with bad cross-country conditions myself.

Farmers often grow flowering crops for seed production: field beans, oil seed rape, strawberries, raspberries as well as top fruit in orchards. These crops benefit by the pollinating services of foraging bees and it is worth the farmer paying the beekeeper a fee to bring his hives near the crop at flowering time, although not many of them are convinced of this. The bees should be taken to the crop the day after the last insecticide spray has been applied and removed before the next. The exercise demands complete co-operation between farmer and beekeeper to avoid harm to the bees. The seed crops always benefit and the bees usually gain something. In good seasons the honey yields from these sources can be considerable.

Whenever bees are transported they are put under stress, and diseases which were latent can flare up with alarming rapidity, leading to the loss of the colony. Frequent long-haul transport should be avoided if at all possible, but short moves cause little trouble with healthy stocks.

When bees locate their hives they note the immediate vicinity, and when they fly afield, the more distant landmarks, up to perhaps 2 km away. If the hive is moved a short distance within that radius a very large number of bees return to the old site using their known landmarks as guides. If you wish to move a hive to a new site 100 metres from the old, then it has to be taken 3 km away for about a week before it is brought back to its new stand. But when the bees are released after a move of over 3 km they relocate rapidly and few, if any, ever get lost.

6

The Four Seasons

Spring

Spring is the testing time for bees. They die out compara-
tively seldom in the winter, but losses occur only too frequently in the spring and early summer, in spite of the
fact that they are often referred to as 'winter losses'. It is
the season of colony growth and expansion in the brood
chamber. Healthy colonies grow at a steadily increasing
pace as new nurse bees emerge to provide milk for the
young larvae and to feed the queen for conversion into
more eggs. If, on the contrary, an infectious disease such
as nosema or amoeba, coupled with dysentry, has left
soiled and infected combs in the brood chamber, more and
more bees become infected and die. Whenever the number of dead bees increases alarmingly, it is as well to have
them checked for adult bee disease (see Chapter 10).

An expanding colony needs a plentiful supply of honey,
pollen and water. Under normal circumstances the over-
wintering stores should last through the spring, but 'heft'
the hives at fortnightly intervals to check their weight.
Hefting consists of lifting up the hives alternately at the
back and front and estimating their comparative weights.
Put 5 to 10 litres of thin syrup (see Chapter 9) on light
hives. Fresh pollen will be coming in from early flowers,
and this can be ascertained by watching the hive entrances
on warm days when the bees are flying and bringing home
the tell-tale loads on their hind legs. In Britain there

is seldom a shortage of spring pollen, but it can happen in some districts in some years; and in many parts of the world, spring pollen starvation seriously retards colony growth. Under such circumstances, feeding pollen substitute made of soya flour and powdered skimmed milk leads to a marked improvement until fresh pollen becomes available.

Bees need water to dilute stored honey, and many diverse water fountains have been devised to supply it without drowning the foragers. Wet sacking, an old sink, plugged and filled with gravel and water, water-lily leaves or duck-weed on a pond can provide moisture, together with the foothold which enables the bees to gather safely the water they require.

Many beginners are tempted to remove the reduced entrance block too soon. A small entrance does little harm at any time and it effectively retards cold spring winds which might chill the brood when there are too few adult bees around. Brood can be chilled and die, or its resistance to disease lowered, if the brood nest is opened in chilly weather or in cold wind. When to open can best be decided by the bees themselves. If you observe them on a sunny day and see a number of young bees indulging in play flights, in addition to the going and coming of the hardier foragers, then it is safe enough to open for the first inspection.

Spring inspection

Opening the hives for spring inspection has a number of definite purposes, it is not just an idle interference. Here is a list of things to look for and do:

1   The brood spread: how many seams are occupied? If seven or eight, it is time to put on a queen excluder and super.
2   Presence of eggs and larvae—these will indicate that the queen is laying.

3 Sealed brood: are the slabs of brood even, flat capped, not too many gaps? Any dark, sunken and perforated cappings should be reported to the Foul Brood Inspector (see Chapter 10).

4 Is there pollen in the ring of cells surrounding the brood?

5 Is there a good store of honey on the outer combs? If not the colony must be fed.

6 This is the right time to look for and mark the queen or clip her wings. If you don't see her, but have seen eggs, you can close down and find her next time.

7 Are there large accumulations of dead bees on the floor? If so they should be cleared away. Lift the brood chambers off the floor, which can be scraped clean with a hive tool, away from the hive stand, before replacing.

8 Are there drone cells or drones?

9 Are there one or two combs available for the queen to lay in?

10 Is this colony less advanced than others in the apiary? If so it may be diseased and a sample should be sent for diagnosis.

11 Excessive burr comb should be removed with a hive tool.

I always carry around, in an envelope, a set of cards, and record some of this information in code. At the head of the card I write the number of the colony, the location, the origin and year of the queen—the top line will read for example:

24   It.   H2   1973

which means: Hive 24—Italian queen bred at home in 1973 from queen no. 2.

Each line down the card then records information at successive inspections, so my next line will read:

4-4-74   Qm—Cl 4/4—Needs super

which means: on 4 April health satisfactory and eggs

seen; a marked queen was clipped on 4 April—eight seams of bees seen, and a super must be put on at next visit. At a later stage the code words: Cx—cells destroyed, QNS—queen not seen, OCL—open cell left, CCL—capped cell left may appear in due season. In this code no record is made of anything that appears satisfactory and normal; only when a positive action has been taken, or when action is required which is not visible from the outside, is a note made. The envelope of cards is taken home and tells me what equipment I shall need on my next visit. At the end of the season, the cards can be filed and new ones made out.

I have indicated that the right time to put on the first super is before the bees have filled the brood nest. Many older beekeepers super far too late and create congestion, which forces swarming on colonies that might otherwise have gone through the season with no trouble. I always advise the use of a queen excluder between the brood and the supers to minimise storage of pollen and breeding in the honey storage combs. In this way, too, the depredations of the wax moth in empty supers is practically eliminated.

Spring management of colonies aims to produce the maximum amount of foraging bees at the time the main honey flow starts. It cannot be too emphatically stressed that strong, populous colonies gather more honey for themselves and the beekeeper than weak ones. During this period we must, at first, exercise some control over colony strength and later some form of swarm prevention, to avoid losing the strength that has been gained. We can discern three peaks in the annual cycle: peak egg-laying, which occurs in spring; the peak in bees which arises three weeks later, when the brood emerges; and the final forager peak, two weeks later still. Thus, working backwards from the honey flow in the area, which we learn by experience, we try to reach brood peak five to six weeks before the main flow and ensure that no food shortage hinders development for some two months before brood peak.

In districts with an early flow at midsummer, progress in brood rearing should be kept steady from three and a half months to one month before the flow, even though unfavourable weather might intervene, and by stimulative feeding when necessary. It is not a bad idea to hang up a calendar in the bee house and mark the weeks when colony growth and food reserves need to be watched. They vary from district to district. Where the flow occurs later (your local association members will help you until you gain experience) it would be most unwise to stimulate early growth of the colony, out of step with the annual floral cycle.

Uniting

Only strong stocks are likely to produce a surplus of honey. Stocks which fail to develop in spring are useless and it is a waste of time to nurse them through, to develop into strong colonies only at the end of the honey flow. If development is retarded by sickness, the bees can be treated early. Colonies that come out of the winter with too few bees to produce the heat they require are unlikely to get ahead in good time. In this case, two weak colonies are best united to make one strong one and produce a strong viable stock.

Bees will unite peaceably on condition that one of the queens is found and destroyed, and that sufficient confusion or delay is caused in the uniting, for the colony odours to become mingled. Two methods of uniting find favour with most experts.

Uniting with newspaper is simple and widely practised, but is not always the best way of achieving its object. You require a sheet of newspaper large enough to cover completely the top of the hive. You then find the queen in one of the hives, kill and discard her, place the sheet of newspaper over the brood chamber and pierce two or three holes in it with a match or hive tool. Next place the queen-right brood chamber on the paper and close up. The following day, the newspaper will be found outside the hive, chewed to a powder, and the bees living amicably together. Bees can be united in this way at any time.

The paper method is less satisfactory when two very weak stocks are united. If there are hardly enough bees to fill half a brood chamber, we cannot expect them suddenly to fill two brood chambers and cope with an unnatural shape of brood nest, split into two separate parts. They can be united by *alternating combs*. We shall need a spare brood chamber. Move one of the hives to one side and put this empty box in its place. Now carry the other hive and place it on the opposite side, so that the empty box is in the middle, between the two stocks you want to unite. First open the weaker colony, using as little smoke as possible, find the queen and kill her; then in the same way find the queen in the second colony and place her with the comb she is on, in the middle of the empty box. From now on use plenty of smoke to confuse the colony odours. Remove, from the sides of both stocks, any combs without brood and shake the adhering bees into the middle box. Then, working from left to right throughout, alternate the brood combs with their adhering bees in the new hive, still using plenty of smoke. Fill up the brood chamber with the best of the removed store combs, bump the now emptied boxes over the top of the frames and give a puff or two of smoke before closing up. The confusion caused by this method of uniting is always successful in spring and summer, but should not be attempted after the main honey flow, when bees are liable to fight and rob.

It will be obvious that, before uniting two weak colonies, definite steps need to be taken to ensure that no obvious disease is present in either. The most frequent cause of weakness, other than sickness, is undoubtedly food shortage, and after uniting the bees must be fed to prevent a recurrence. Sometimes ageing queens are the cause and, after uniting, a young, vigorous queen may have to be introduced. In addition, late swarms that cannot individually 'set up house' in time for the winter will be able, jointly, to achieve what they could not do in separate hives. If these conditions do not apply, then suspect the presence of some disease and act accordingly.

Summer

Spring passes into summer, and the brood box is still filled with brood and packed with bees; honey is being stored in the supers above. The queen may not be able to expand beyond a certain point when the number of young bees emerging each day balances her egg-laying rate. At some stage near this point, with potent drones in the hive, she will lay an egg in each of the queen cups. Three days later the egg hatches and the larva floats on a bed of royal jelly food; five or six days later still, the old queen will fly out with a swarm composed of half the bees. From late spring until mid-summer, the hive must be opened and 'inspected' to see whether there are cells in the hive, at intervals of nine days or less.

Summer is the time of the honey flow, but alas seasons vary, and even old hands sometimes fail to realise the danger of a cold spell in early summer. With so many bees around, food is consumed very rapidly indeed. The English summers of 1972 and 1975 caught many beekeepers unawares. Those who did not feed their bees in June and the first week of July found them dead or depleted, but those who did feed reaped a creditable harvest in a few weeks of glorious weather.

Keep well ahead of the bees in putting on supers. Even if they never fill them, a strong colony in a National or Smith hive needs four supers quite early in the summer just to accommodate the work force; in a Langstroth, three supers are required, and in a Dadant or Jumbo, two supers. It may appear paradoxical that in poor seasons the bees should need these supers more than in the good; but under these circumstances, even the foragers have to stay home all day and need room to avoid congestion on the brood combs. This lesson has not been fully appreciated by many beekeepers, who regard supers solely as storage space for their honey crop and overlook the essential needs of their bees.

The honey flow

Whilst the main honey flow comes in the summer in our temperate climate, it may be all over and finished in Mediterranean and sub-tropical regions. But even in the narrow confines of one small country, differences in the flora, the situation of the apiary on a slope or in a valley, or even differences in the composition of the soil and annual weather variations, make it impossible to help the new beekeeper with any practical scheme for dating the flow. I have seen many practical treatises on beekeeping which give graphs showing the dates of the flows, dates for supering and dates for swarming. The unfortunate reader who follows such guidance is bound to meet with trouble, because the annual rhythm of bees and plants responds to external factors which broadly follow the seasons but cannot be confined to dates on a calendar.

A beginner may well wonder what constitutes a honey flow, but as soon as it arrives no doubts will remain. A constant stream of bees tumbles out of the hive to be met by an eqally strong contingent of returning foragers. The hum of flying bees fills the air. The bees can be handled with the utmost ease, intent only on bringing home the plentiful nectar, processing it and sealing it over. Honey and pollen, the main food reserves in the colony, are brought in and stored against the winter. All the flying bees take part; even the feeding of the queen takes second place and her breeding declines. The brood nest shrinks and every available cell on the outside of the dwindling brood patch is used for storing food.

And suddenly the flow stops. All the good temper which

accompanied the gathering in of the nectar ceases, the bees turn all their attention to defending their gains and are unapproachable. Heed their warning and leave the brood nest well alone. It will soon be time to harvest the crop and make preparations for the winter. Throughout the spring and summer we have not looked at the combs in the supers—indeed, the frames are not designed to be

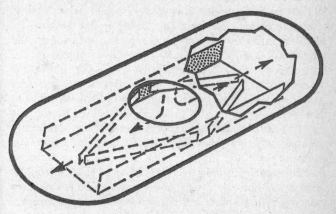

FIG. 20   The Porter escape

separated in this way. Leave them on for a week or two after the main flow, then put the Porter escapes on the clearer board and place it above the excluder under the supers. If large roof ventilators are used, lay a sheet of newspaper between the roof and the uppermost super before replacing, to keep light from the ventilators out of the supers, and leave for 24 hours for the bees to clear. Remove the supers to the honey house, shed or kitchen and make absolutely sure that the stack is completely bee-proof. Sheets of hardboard, cut to size and weighted down, can be fitted over the top of a pile of supers while they await extraction. If you are not 'going to the heather' your season is finished, and you remove the queen excluder and the Porter escapes, and put in the reduced entrance blocks; then leave the bees to gather whatever they can from late flowers.

In parts of Western Europe and Britain there are often ling-heather moors within 50 miles, and at that distance it makes good economic sense to take all the hives to the

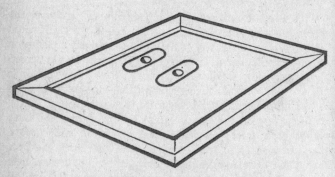

FIG. 21   The BSS inner cover and clearer board

heather to obtain their winter feed. In the North and in some parts of the South and South-west of England, in Scotland, parts of Ireland, Brittany and the German moors, ling provides the main flow. Beekeepers in areas nearby bring their bees to the moors as soon as the ling starts to flower. Their aim is to have their colonies stock up the brood chamber solid with honey and pollen for the winter.

It is often said that heather honey is a poor food for the bees to winter on. That may be the case in districts with hard, prolonged continental winters, but I have found the exact opposite in my area of Southern England. On the contrary, those colonies which winter with a full brood chamber of heather honey and pollen invariably get through better than any others. This is probably because in milder climates bees can make cleansing flights at frequent intervals. Some hives with stocks specially raised to reach their peak, gathering force when the ling blooms, can produce a surplus of this delicious honey. For a good heather surplus it is essential to bring the bees to the same condition as any colony prior to the main flow.

Transporting hives to the crop

When hives are moved during the summer, whatever the crop, the bees need ample ventilation and plenty of room to move away from the brood. The hive parts are stapled together to prevent them breaking apart. And a screen board must be made or obtained for each hive. This is like a cover board but has perforated zinc or wire cloth covering the central area instead of wooden boards.

First, remove the honey supers by means of the clearer board as indicated above, but leave an empty super above the excluder. The super stays on the hive until the bees are ready to return home. Before venturing further, heft the hive to check that there are sufficient stores in case the flow does not materialise for some weeks. If you think the stores are insufficient, take some syrup and a feeder with you to leave the bees a feed after they have arrived at the other end. Then put the screen over the super in place of the crown board. The remainder of the process has to be performed quite late, when the bees are all home for the night. Insert the entrance block in the closed position or block the entrance with a strip of foam plastic, 2.5 cm square and 2 cm longer than the entrance block. Smoke may be needed to drive back the guards at the entrance or even a cluster of bees hanging out on a warm night. Close all the hives before attempting to nail up any of them. With the entrance blocked, tack the screen on to the hive sides with four nails and staple the floor to the brood box with four hive staples (from appliance dealers). Two staples are driven across the join on two opposite sides of the hive. Then staple across the join between the brood chamber and super, bridging the queen excluder. Stand the hive inside the upturned roof and all is now ready to go, but with some note of warning. If the hives are to travel in the boot of a car, fix the boot cover to the best of your ability so that it does not lock down hermetically; if a large number of hives are transported in a van, ventilation of the compartment is necessary, and if hives are stood one on top of the other lay some pieces of 5 cm × 5 cm timber between layers to keep the screens open.

As soon as the hives arrive at the new location, lift them out of the roofs, place them in their final positions, put on the roofs and then walk along the row, opening the entrances 3 cm or so. If you linger over the opening of the first one you are unlikely to make the same mistake with the others. I always prefer to move bees at night and creep home in the small hours to bed. On the few occasions when I started early in the morning, the heat of the day caused me great anxiety before the bees were liberated, but they did not suffer unduly. The heat generated by a strong stock of bees, closed up with inadequate ventilation, melts the wax combs and honey in a very short space of time and the stock perishes in the sticky mess.

Some beekeepers prefer to strap their hives together with metal or plastic bands. Best results are obtained when the bands are tautened with specially designed lever machines. These are somewhat expensive for just a few hives, but invaluable for those who move bees several times a year. Quite useful fibre bands complete with toggle fasteners are now on the market and function satisfactorily. Band fasteners avoid the constant piercing and eventual damage of hive sides, but they must be really tight to resist movement between the hive parts during transport. The use of ropes for tying hives together often leads to disaster, because they have too much stretch to hold a sudden strain, such as might arise during an emergency stop.

One super is enough to carry and enough for a heather crop, but when bees are moved to gather a crop of oil-seed rape, field beans or clover in the height of the summer, a second super may be needed after the bees have been opened up.

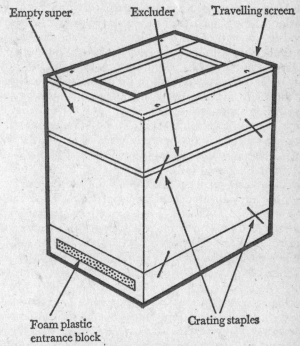

Empty super   Excluder   Travelling screen

Foam plastic entrance block

Crating staples

FIG. 22   Hive crated for transporting

Autumn

When the crop and the excluders have been removed at the end of the summer, and the narrow entrance block has been inserted, hefting will soon reveal whether the brood chamber contains sufficient stores. I think it is just as well to leave them well alone for a week or two. Just check that the drones are being disposed of by watching the entrances. Remember that opening the hives will only encourage robbing. If you see any unusual commotion outside a hive, with dead bees around the entrance, you can be reasonably sure the hive is being robbed out. Put in the entrance block at the contracted position to enable the guards to gain control. It may even be necessary to put a forkful of hay, straw or weeds over the entrance to impede the robbers. Check for any other possible leaks in the hive that bees can get through and block them up with a plug of grass or plasticine. Robbing is better avoided, by leaving no bits of comb, honey or wax around, than cured once it has started.

About a month after the end of the main honey flow, or immediately after the heather, the bees once again become amenable. It is time to check for stores. The brood chamber needs to be full of honey and pollen. If it is not full, feed until it is, preferably in the evening. Bees do not overfeed; they store food and use it as required. When you feed, use the entrance blocks in the restricted position.

Storing supers

The honey stored in the supers can be looked on as the beekeeper's just reward for his efforts. Much of it will be extracted, as we shall see in Chapter 8, and the shallow supers, still wet with honey, have to be carefully dried and preserved safe from wax moth, damp and mice, ready for the following season. The bees will clean up the combs and strengthen the edges of the cells if the supers are replaced on the hives for a week or two, above the cover board with the roof directly on top of them. Of course, when they are on, the entrance should be restricted, if this has not already been done. The supers are removed quietly at night when the bees are down below, without recourse to a clearer board.

Store the supers stacked in a dry shed, or even outside with a spare roof over the pile, but they must be made mouse-proof. I always place a queen excluder at the bottom of the stack and adjust the successive supers carefully to leave no gaps. Even when the stack is built up inside a shed I put on some kind of cover to keep dust out of the combs. Later, during the winter, they should be taken down one by one and the outside of the frames scraped clear of wax and propolis. Some experts prefer to store combs wet, as they come from the extractor, but I dislike this habit because bees constantly try to rob the combs and the residual honey damps off and ferments. The cells, too, are liable to damage since the edges have not been thickened and strengthened, as they are when given back for clearing.

Winter

Breeding usually stops in the autumn and early winter. As the nights get cooler and bees draw closer together in a cluster. In some ways the cluster has the shape of a swarm, but it is split in horizontal slabs by the combs, with bees filling all the empty cells and the seams to generate and save heat. The colder the weather, the tighter the cluster. Some honey is consumed and the cluster wanders across the combs very slowly as they warm up the stores. At the very coldest part of the year, the cluster has drawn in tight enough for the temperature at the centre to reach 31 to 33°C, and the queen can resume her laying. In the northern hemisphere laying usually starts near the turn of the year.

It has now become clear that bees need moderately cold weather in winter, and in temperate climates it is a great mistake to pack the hives with outer cover, sawdust wraps, glass wool or other forms of wrapping. These may indeed be necessary in Central and Eastern Europe and the Northern States of the American continent where the severity of the winters indicate that bees would otherwise die. Investigations show that there is very little difference on the outside of the cluster between ambient temperature inside and outside the hive, whatever the packing, and further research has shown that colonies kept artificially warm in winter fare no better, and sometimes worse, than when they follow the natural seasonal rhythm of warm summer and cold winter. The best packing for bees is bees and sealed stores. The more of both the better. A 4-cm layer of slowly moving air above the cover board provides the best form of insulation over the cluster.

The native race of bee, or bees, acclimatised to an area over several generations, copes better with the flora and seasons than bees imported from other climates. Temperature control in the winter cluster depends on the availability of food, the numbers of bees in the cluster, and the avoidance of disturbing elements such as damp, wind, woodpeckers, mice and accumulation of wastes. The beekeeper ensures that food and shelter are adequate; the bees see to the rest, provided that they can fly on fine warm days.

Cleansing flights

Bees do not normally defecate in the hive. In cold weather, waste products accumulate at the end of the lower gut in a sac known as the rectum. If their food has undergone no fermentation and contains only a modest amount of non-digestible residue, the faecal matter swells the rectum and can be retained for several weeks and voided when the weather permits a cleansing flight. When the winter food is unsuitable, dysentry may appear and the combs become soiled with streaks of dark brown faeces, due to the bees' inability to retain the additional waste matter.

Waterproofing

It is evident that the roof should be waterproof, to keep out rain, and all commercially produced hives meet this requirement. Less evident is the need for a roof resistant to wind. A roof which lifts in strong wind and gets blown right off the hive must be avoided. Roofs should have sides deep enough to wedge against the hive when a strong wind raises one side, and this is obtained when they fit reasonably tight. A minimum depth of 15 cm and a clearance all round of approximately 5 mm gives enough latitude for lifting and replacing the roof, as well as safety in winter.

An aspect of waterproofing which is often lost sight of is the tilt of the hive on the stand. Small though the restricted entrance in the winter block may be, it is still enough to allow driving rain to enter and accumulate on the back of the floor if it slopes towards the rear. When setting up the stands, make sure that there is a slight tilt towards the entrance so that water can drain from the front.

Ventilation

Winter ventilation is just as much part of waterproofing a hive as the provision of a sound roof. Bees in the cluster breathe in air and exhale carbon dioxide and water vapour. Unless there is some means of removing these wastes, the water will condense on the cold hive walls and the inmates will lack oxygen. But the draught required should not be forced through the centre of the cluster. If the standard crown board with its central feed hole is used, then obviously air, heated by the bees, will rise directly to the central hole and find its way out through the roof ventilators, to be replaced by cold air flowing from outside and through the cluster. The movement of air causes a cold through-draught at all times by convection, and failure to understand what is happening has led some beekeepers to abandon winter ventilation altogether.

A better solution can be found, however, by covering up the normal escape or feed holes with a piece of glass or wood and raising the cover board about 3 mm. The operation should be carried out quite late in the season, as soon as active flying has ceased, by placing four match-sticks or similar wooden sticks across the four corners of the brood chamber and replacing the crown board over them. If action is taken too soon, the bees propolise the gap before the hard weather sets in; but, correctly carried out, this simple operation provides peripheral ventilation round the cluster and reduces the flow of cold air to a minimum. At the same time, warm gases can rise up through the frames and carry off the waste. The matches can be removed in spring when the bees begin to fly freely. Top ventilation is a winter problem only: the bees see to aeration of the hive when the cluster breaks up in spring.

The 'New Standard' cover board

Those who wish to ensure adequate perimeter ventilation and an insulating air cushion above the frames could well consider making the crown board advocated by E. J. Tred-well, former Beekeeping Lecturer at the Hampshire College of Agriculture, and keeping the standard crown board for use as a clearer and feeder board. This is quite a good arrangement, because at best the standard article is a compromise.

The 'New Standard' crown board can be made quite simply by cutting a piece of 6-mm waterproof or marine plywood to the exact outside dimensions of the hive. Mark off a line parallel with and 5 cm from the sides; then mark off 5-cm centres all round this line and drill 16-mm holes through the marks. Four pieces of 22-mm square timber cut to fit round the outside, nailed and glued, complete the cover for hives with top bee-space. But for the National and Commercial hives, four pieces of 22 mm × 6 mm lath are similarly glued to the underside to ensure a bee-space over the frames. With these crown boards, match sticks are not necessary and there is no central hole to cover over. The 22-mm deep rim provides an insulating air cushion above the cluster.

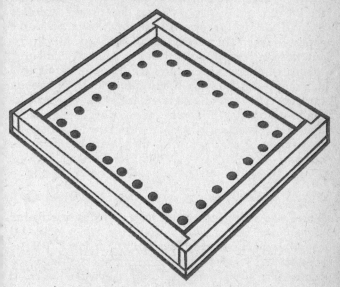

FIG. 23 New Standard inner cover

Winter entrance

The modern hive has a simple wooden square block which is pushed into the front, between the floor and the brood chamber, and comes to rest against pins or a rebate in the side to prevent it from sliding right in. It has a gap cut in one side, 6 mm deep and about 75 mm long. For closing the hive, the gap faces forward and the block completely closes the entrance, but when the slot faces upwards it allows a contracted entrance for the bees. For some unaccountable reason, appliance dealers' catalogues and many books, including the Ministry pamphlet, show this narrow gap below the block instead of above it. With en-

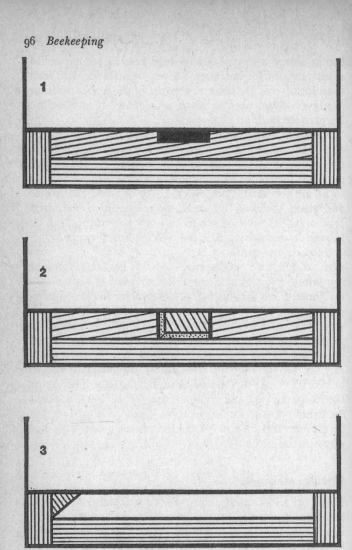

FIG. 24   Entrance block in use. 1 restricted entrance. 2 closed for transport. 3 removed for summer; stored on cover board.

trance below, a 6-mm layer of dead bees in winter would effectively seal the entrance, prevent ventilation and lead to suffocation of the colony, whereas in the upper position the layer would need to reach a height of 16 mm before harm could be done.

Some entrance blocks come with two different lengths of gap cut on adjoining sides. This is a great pity, because the double cutting weakens the block at the centre and makes it ineffective as a complete closure. It is advisable to fill up the smaller gap with a slip of wood cut to size and glued in flush. If a smaller gap should be required temporarily, a short length of lath or a tuft of grass can be pushed in one side, but I have never found it necessary.

The reduced entrance serves a double purpose: from early winter until mid-spring it holds back the force of the wind, and prevents the entry of mice. In some old-fashioned hives various bits of perforated metal or queen excluder are tacked across the entrance to perform a job which the entrance block does quite simply. When in late spring the block is removed, it is kept on the hive crown board, under the roof to avoid losing it. Placed diagonally, it will fit even on the square National hive; the height of the crown board framing and the plinths or blocks on the roof will accommodate its thickness without disturbing the seating of the roof. When feeding, it is put in the entrance in the contracted position and does not obstruct the feed hole.

Cleaning equipment

Clean up all the spare equipment during the winter, to have it ready for use during the spring and summer. There is a good practical reason why apiary equipment should be dealt with in cold weather. Cold makes both wax and propolis hard and brittle and a hive tool or paint scraper will remove them quickly. I clean up hives and frames over a tea chest so that the dislodged particles can fall directly into it ready for melting down. For scraping frames a couple of boards spanning the tea chest support the lugs. Avoid breaking or damaging super combs, but scrape wax and propolis from side-bars, top-bars and bottom-bars and replace them in the scraped-out super. The boxes need to be checked over too, for weakness in the joints, damaged wood, shrinkage and knot holes, all of which can be repaired. It is a good plan to scrape and scrub the spare floors and hive bodies with hot water, soap or soda and a stiff scrubbing brush, and set them out to drain and dry before stacking them away.

Old brood combs that require replacing because they are misshapen or have too many drone cells are placed flat on the boards over the tea chest; a hive tool is pushed down through the comb near the frame to break the wires, and the whole comb is dropped into the wax-chest. Carefully supporting the frame underneath, scrape the sides and the lugs, not forgetting the lug ends; then remove the wedge from under the top bar, using the hive tool as a lever, and scrape all the old wax away. If frames come from a colony which suffered from a disease they will need to be disinfected, as explained in Chapter 10. They are then ready to receive a new sheet of foundation.

With the cleaning up done, the feeders washed in hot water, the smoker decoked, you can have some confidence that next year will prove to be better than the last and make plans, order any new equipment in good time, and read and learn. There is no point at which you can rest and say you know all about bees and beekeeping.

7

Swarms and Swarm Prevention

Before we can hope to deal with swarms, we must understand the nature and causes of swarming, at least in so far as research has been able to explain the phenomenon.

The unit of bee life is the colony, not the individual bee, and therefore, if the species is to continue at all, the colony must reproduce itself. Honeybees have evolved swarming as their answer to this problem. Swarming must take place early enough in the annual cycle for the swarm to survive, establish a home, build comb for the brood and store against the winter. Swarming seldom if ever takes place after the main honey flow, because starvation would inevitably follow during the ensuing winter. Other social insects have found different answers, but honeybees are entirely dependent for their annual food supplies on the relatively short period when flowers bloom in profusion. They abandon their old hive with its food reserves, combs, brood and queen cells, and set up a new home elsewhere, with only the honey they carry in their crops. Since they leave with their 'old queen' she quickly resumes laying, and there are plenty of young bees in the swarm in condition to nurse the first batches of larvae to maturity. The colony they leave behind has stores and bees enough to survive the weeks of waiting required before the new queen begins to lay.

Swarming, then, by its very nature means an increase, by division, in the number of colonies, and the division does not always stop at the first swarm. A week after the first or prime swarm has left the hive, several queens may emerge from their cells. There will be some deaths from desultory fighting amongst them but, weather permitting, a newly emerged virgin may well fly off, taking half the remaining bees with her; and in quick succession, often on the same day, further secondary swarms or casts may leave, each taking about half the residual bees. Such casts are often quite small and their chances of survival become less as they diminish in size, leaving the beekeeper with a very small unit that will need all the surplus honey gathered in the preceding weeks to re-establish itself with sufficient strength to last the winter. Clearly, if we are to earn the name of bee *keepers*, we shall have to learn how to keep our bees, and not just to stand by and lose them.

The obvious remedy would be to approach the problem by breeding a non-swarming race of bees, but unfortunately their nature prevents such a solution being final. Queen bees mate in the air at some distance from the hive, with drones from unknown sources, so that even if we did manage to breed a non-swarming bee, it is unlikely that the strain would last very long in our apiaries. As soon as we try to increase or replace failing queens by natural or artificial means, cross-breeding occurs and the strain is lost. Artificial insemination of queens to produce desirable strains is being successfully carried out, but the micro-techniques involved are not likely ever to lead to its widespread adoption. We are therefore left, until Utopia arrives, with the annual tussle between the bees who want to divide and the beekeeper who must maintain his foraging force in a single unit.

To some extent, we can help ourselves by avoiding activities that provoke swarming. Some bees are more prone to swarming than others. We suffer, unfortunately, from the sins of our fathers, whose methods of beekeeping encouraged swarms and entailed destruction of the bees every autumn to get the honey. They kept over to the next

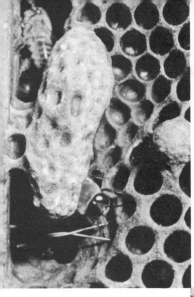

PLATE 1 Three kinds of cells in a hive. Worker at top right, drone bottom right and sealed queen cell on the left

PLATE 2 A 'wild' colony in a roof. Failure to prevent swarms from escaping can cause a nuisance to neighbours and creates unwanted competition for forage. The centre combs were 65 cm long and 35 cm deep

PLATE 3　　Adult drone and workers

PLATE 4　　'Court' behaviour round a queen, as workers lick queen-substance from her body. The queen is busy laying eggs

PLATE 5 Queen on a slab of sealed brood. Workers often rush to cover sensitive larvae or to the honey stores when combs are lifted out, leaving the queen to seek shelter as best she can

PLATE 6 Bee foraging on a clover flower with tongue fully extended

PLATE 7   Bees taking syrup from a polythene bag, placed on the
inner cover, under a bee-tight roof

PLATE 8   Rendering old wax combs at low cost. Scrap wood heats
the pan. Molten wax is ladled from the centre of a conical strainer

year the medium-weight hives, which were either prime swarms or their parent colonies. They killed the heavy colonies—those which had not swarmed—and the light ones—the late swarms or casts. So, after centuries of selection, the non-swarming strains were weeded out and only those which swarmed annually were maintained. Since then, importation of bees from countries, mainly Italy, where this selection process was less practised, has helped to redress the balance; but for reasons of climate and the prolificacy of the immigrants, the position has changed but little.

Swarming serves the double purpose of replacing the old queen by a new one and of colony reproduction. Some strains of bee swarm every year and even with non-swarming strains, weather conditions may upset the pattern—with the result that in some areas, North-West Europe and Britain in particular, every colony has to be treated as a potential swarmer every year, at some unknown time between early spring and the end of the main flow. In other, more reliable climates, the swarming season is considerably restricted and control measures are required over a period of little more than a month each year and at more predictable times.

Queen replacement by supersedure

Some strains of bee tend to replace their failing queens by a process known as 'supersedure'. This usually takes place in late summer, after the honey flow, when a swarm would be unlikely to survive. A few queen cells appear and a young queen emerges, mates and starts laying. Sometimes the old queen is killed at once, but occasionally she survives for several weeks, laying in the same nest as her daughter. This trait is most desirable but alas may soon disappear through cross-breeding of the daughters. Supersedure can occur early in the year and many colonies seem to start the supersedure process, but the potential for the

kind of imbalance which triggers off swarming is so great
that they also end up by swarming.

Queen replacement by swarming

The balance in a stable bee society is largely maintained
by the distribution of the queen substance pheromones,
which inhibit cell building and queen raising. If either
the supply or the distribution of queen substance is curtailed, the colony is likely to prepare to swarm, and once
the process has started it is very difficult to halt it. A queen
appears to produce less pheromone at the very time when
it is most needed: at the period of peak brood. An old
queen may produce less than a young one and the colony
she heads is more likely to swarm. This can also happen to
a damaged, sick or introduced queen. Proper distribution
of the available pheromone is prevented by congestion of
adult bees on the combs. Young bees show great reluctance to leave the brood and always congregate most thickly
over the areas with large patches of unsealed brood. If
the capacity of the hive is too small to contain comfortably
all the bees, at night or in poor weather, congestion results in uneven distribution of the inhibiting substance.
Some relief is obtained when supers are added to increase
the space available and to reduce pressure over the brood
area, but when vast numbers of young bees start emerging,
congestion can still occur in the centre of the nest, simply
because nurse bees will not move away from the brood.
Some improvement can be obtained by inserting, outside
the nest, one or two frames of foundation to be drawn out
by the wax-producing workers, which are thus artificially
separated from the nurse bees. This point should be borne
in mind and used as part of our management system, especially when we keep bees in such small hives as the
National or Smith.

Usually, eight days after the eggs have been laid in
them, the queen cells are capped over and the old queen

flies off with approximately half the bees; before they go, the bees fill their crops with nectar and honey from the open cells, and perform a special whirring dance on the combs. For an hour before they leave, vast numbers of workers cluster in serried ranks on the combs, with their heads pointing upwards. I have often watched experienced beekeepers at demonstrations open a hive and immediately announce to their audience 'those bees will be swarming within half an hour' and close the hive. Sure enough, the bees come out, inside the allotted time. The swarm leaves the hive and swirls around in the air for a few minutes before the bees 'pitch' on a convenient spot, where they all cluster. The place chosen for pitching may be convenient for the bees but much less so for the beekeeper who is called upon to remove them. Scout bees are sent out to search for a likely cavity for the future home and report back to the cluster by 'dancing' to indicate the location of the chosen residence. After some hours or days the whole swarm takes off and goes straight to the new abode; the bees cluster again to produce wax as material for the combs, while those foragers too old to produce wax set out to gather pollen and nectar for provisioning the new nest. At least three weeks must elapse before the first eggs hatch, so during that period natural wastage will inevitably lead to a reduction in the swarm's population. The old queen is normally superseded some weeks later.

The parent colony has now lost its queen and half its bees and no new eggs can be laid until one of the queens emerges from its cell a week later, goes out, finds a mate and resumes laying. During the week following the emergence of the prime swarm, the colony's numbers will be swelled by newly emerged bees and, unless prevented, many of these may be lost in subsequent casts. Casts behave like swarms except that, after establishing themselves, the virgin queens have to go out and mate. Their development is therefore somewhat delayed, but the wastage by deaths tends to be smaller in casts, because of the greater proportion of young bees which do not begin to age until breeding resumes, when the swarming fever is over. The parent colony, still being strengthened by emerging bees, only starts expanding when the new queen is in full lay, some four to six weeks after the first swarm.

Is swarm control necessary?

This summary of the causes and nature of swarming leads us to examine its effects on the beekeeper and others. On the basis of the annual or biennial increase, if we caught and retained the swarm our beekeeping would expand in geometric progression, and if we started with only two hives we should soon find ourselves in possession of dozens or hundreds of colonies. This would involve considerable capital expenditure, total consecration to beekeeping and very little, if any, return in terms of honey or cash. Our honey harvest increases in proportion to the foragers available at the time of the main honey flow and, as we have seen, reduction of half the colony's strength, precisely at the beginning of the flow, results in a disproportionate reduction in our crop of surplus honey. Each colony requires a considerable amount of honey-sugars for its own maintenance—some estimate it at 100 kg per annum—and any surplus the beekeeper may reap must be in addition to the bees' own requirements. If the stock is split in two at the beginning of the flow, both colonies will require a separate maintenance ration and the bees available for collecting it will be reduced by half in each of the new units. The advantage of having two queens producing double the amount of workers is more apparent than real because the new force of gatherers arrives on the scene after, rather than during, the flow. It follows that swarming has to be controlled in some way to save the crop.

In our modern world we live surrounded by neighbours whose understanding of bees is minimal and their liking for them even less. We have social responsibilities to see that our management of the bees causes no inconvenience, real or imaginary, to those who surround us. The arrival of a swirling swarm of tens of thousands of bees alighting from over the fence on a tree branch just above baby's pram constitutes for them a real danger, even though we may know that it exists only in their imagination. If we did not see the swarm leave and it eventually settles in their wall cavity or under their roof, they will be put to some expense to have it removed; if they do not remove it, the stray swarm competes with our own bees for the available forage. A number of wild colonies in a district can seriously reduce our own takes of honey. I remember once visiting an apiary where over a dozen wild colonies had taken up residence in the cavity behind the wood cladding of a large outhouse, and there were signs that more had been there at some previous time. Twelve wild colonies severely restricts the possible size of an apiary. We must therefore consider it our duty to the non-beekeepers around us, as well as in our own interest, to so manage our bees that they do not swarm, or if they do, to at least take steps to retrieve the swarms.

Taking swarms

Bees in a swarm cluster have just taken their fill of honey and seldom sting until the reserves in their crops are used up. When this happens, usually after about 48 hours, the 'stale' swarm, without reserves, begins to consider the cluster as its new home and is often found to have some drawn comb at its centre. This they defend as they would a hive. The swarm's only chance of survival is with the queen, and unless the queen is taken with the swarm all the bees will return to her. To take a swarm you will need a veil, at least to begin with, a straw skep or a cardboard box about 30 cm cube, and a lighted smoker, although it will only be needed in emergencies: never puff smoke directly into a swarm.

If the swarm has pitched on a bush, a pair of secateurs can be used to cut the branch and free the twigs on which it hangs, leaving enough stem to hold on to while trimming, and the branch is just carried over to the open skep and sharply shaken into it; the skep is then inverted and raised an inch or so at one end with a stone or brick. It is best to place the skep in the shade and leave it until evening before hiving. If the swarm is hung on a tree branch, hold the skep under the branch and shake it sharply once or twice to dislodge the cluster.

Bees pitch in many inconvenient and awkward places, and it would be futile to try to enumerate them. A little experience with a beekeeper will soon teach the basic facts of how to deal with difficult places, but I will endeavour to set out the basic principles to follow. An awkward place in this context means one where the whole swarm, with the queen, cannot be dislodged at one and the same time. In all these circumstances an additional skep or box will be needed as well as a bee brush, which anyone can make by using a sheet of 9-mm foam plastic, about 20 cm × 7.5 cm, centrally between two pieces of 25-mm lath, long enough to leave a handle, the floppy edges providing the best bee brush imaginable. A handful of long grass serves in an emergency. The other essential item is a queen cage; the most easily obtainable variety is a plastic or metal tubular hair roller with a cork at each end, but make sure the mesh is small enough to retain the queen. Place the skep under the swarm and brush as many bees as possible into it. Look into the open box to see if you can see the queen. If the bees scatter or clump they can be bumped to one side. As soon as the queen is seen, pick her up by the wings or thorax, and place her in the queen cage; close the cage and fix it with a short length of thin wire in the top of the skep; invert the skep with one edge resting on a brick and knock the bees onto the ground in front of the skep containing the queen. It will not be long before the bees find her and call in the others by fanning across their scent glands.

If, on the other hand, the queen is not found in the first batch of bees, set an empty skep inverted on a brick and throw a few of the bees at the entrance and the rest about 1 metre away. As they run to join the others they string out and the queen can be readily picked out. If she is still not with them, it will become obvious by this time because they will be lifting off, trying to rejoin the original cluster. Then is the time to go back and remove some more bees, repeating the process until the queen is found, each time adding the small batches to the second skep, after examining each for the queen. The bees seldom show any inclination to sting if they are handled or brushed gently and steadily. With practice, the queen will be found easily enough. Many beekeepers trust to luck that the queen is in the skep; she most often is, but does not always stay there. When the queen is captured and caged, the swarm cannot leave. Stray bees will continue to fly around until evening, when the swarm can be hived. Place a large piece of sacking or muslin on the ground, lift the skep and place it gently in the centre of the sacking. Pull up the edges and corners and tie them around the skep with a piece of string to carry home. Beekeepers who have a nucleus hive can run their swarm direct into it, with the queen cage suspended between the combs. The entrance is closed with a piece of perforated metal strip and the cover secured for transport.

Preventing casts

The fact that a swarm has left one of your hives represents a breakdown in your system of management which you now have to do your best to correct. While the swarm is settling, waiting for its evening removal, go through the hives to ascertain which one has swarmed. This task is simple enough since the hive concerned will have many less adult bees on the combs and will certainly have one or more sealed queen cells, and probably several open ones as well. At this inspection, choose a good open cell on a comb and insert a drawing pin on the top bar just above it; gently brush off adhering bees and remove or pinch any other cells on the comb. Always choose your cell before destroying any. Shake the remaining combs clear of bees and destroy all other cells; then close up the old hive and prepare a new one, floor, brood chamber with frames and foundation, cover board and roof, which you can stand for the time being at the side of the parent stock. Prepare a hive stand a few yards away and facing in a slightly different direction from the parent. That stand will receive the parent colony, not the swarm.

Hiving the swarm

When the swarm is brought back in the evening, first remove the queen excluder, supers and roof from the parent colony, in one unit if possible, but without disturbing the bees, and place it on the upturned roof of the new hive. Cover the parent colony with a temporary cloth or cover board and carry the whole brood chamber onto the prepared stand. Put the new hive on the stand where the parent colony stood before; remove three frames of foundation from one side of the hive and untie the string round the skep. If the queen is caged, remove the cage very gently. Have no fear of putting your hand into the swarm, but move very slowly and deliberately without jerking. Then shake the bees into the gap in the frames and slip on the cover board gently, so as not to crush bees. Go to the front of the hive where some of the bees will already have begun fanning and open one end of the cage; put the open end in the hive entrance and let the queen walk in to rejoin the swarm. Now remove the cover board quickly, but gently, put on the excluder and supers with cover board and roof. Go over to the parent colony and cover that one up too. About half an hour later the supers can be pushed over to one side and the removed frames slid into position one by one without disturbing the colony. If the swarm was run into a nucleus box, simply put the frames, with bees, into the centre of the new hive, fill up with frames and release the queen.

Some beekeepers prefer to 'run' the swarm into a new hive rather than throw it in at the top. For this a hiving board will be needed, as wide as the hive and long enough to reach from the hive entrance to the ground in a gentle slope, not above 30° gradient. With the board firmly fixed and the old brood chamber removed to its new position, the hive is placed on the stand with queen excluder and supers in position, and the swarm is shaken onto the hiving board. As soon as the bees start running up, the queen can be released at the entrance, as above. Of course, if the queen has not been caged, whichever system is preferred, the bees are just shaken out together with the queen.

Under the conditions just described, the swarm can develop rapidly, whatever the weather, by drawing on the stores in the supers. But in every case where a swarm is hived without such reserves, the beekeeper's first duty is to feed 5 litres of syrup. The food provides for natural expansion and enables a late swarm to reach full strength before winter, and an early one to produce a small surplus.

Reduced to their simplest form the steps can be summarised as follows:

1 Catch the swarm.
2 Move the old colony's brood chamber to a new site, removing all queen cells but one.
3 Hive the swarm in a new hive on the old site adding the queen excluder and supers from the parent colony.
4 Feed, if there are no reserves in the supers.

In this way the swarm retains the surplus already gained, keeps all the bees that may have been in supers and, for some days afterwards, will continue to gain, from the parent stock, those foragers which return to the site of their old home. In this way, too, the swarm will produce surplus honey in spite of the dislocation. There will be less than if there had been no division, but the loss is considerably reduced.

Increase by natural swarming, nevertheless, results from a degree of neglect or ignorance on the part of the bee-keeper, who should learn to cope in a more skilful way with his bees' proclivity to divide before the honey flow.

Management to delay swarming

During the entire period when our stocks are likely to swarm we examine them every nine days at least, but for the amateur a weekly examination has to be the norm. The inspection has definite purposes which must be kept in mind every time a hive is opened. These are to check:

1 Presence of eggs and queen.
2 Sufficiency of stores.
3 Room for expansion of brood area.
4 Presence of queen cells.

The first and second points have already been dealt with, but the third and fourth relate directly to what is commonly known as 'swarm control'.

Room for expansion is already built into the larger hives and needs no further effort on the part of the bee-keeper, but how can extra room be provided in a brood chamber of limited size that is already filled with comb? Some beekeepers advocate doubling the brood chambers either with a full depth brood body or with a shallow box. This latter '1½ system' has no advantages at all, and does not even enjoy the facility of comb interchange between the two sections of brood chamber. The brood nest is artificially split into two throughout the year, with the gap at the precise point which cuts through the middle of the nest during the critical spring period, when the cluster needs to be kept as compact as possible. The use of two equal brood chambers has less disadvantages, and is widely practised in countries where prolific bees and prolonged flows make less demands on the beekeeper for swarm control. In Britain, however, one can be permitted some apprehension at the prospect of looking for swarm cells through two brood boxes at every inspection, or of chasing a queen over twenty or twenty-two combs when it becomes necessary to find her. One of the main advantages claimed for the double brood chamber is the ease of checking for cells at the routine inspections. The upper brood chamber is raised at the back, drawn over the lower box and tilted to rest at right angles while a search is made for cells under the upper combs and above the lower. After trying this system for many years, I gave it up because several times the bees built cells too high in the upper box or too low in the lower box to be seen. Inconsistencies of this kind nullify all the effort put into regular swarm control. The very existence of these methods points inevitably to the inadequate size of the brood chamber in use.

Increasing the brood area

I commonly practise two methods with the small National and Smith hives, and sometimes the Langstroth: evening up the stocks, and using deep supers. When inspecting the apiary, some stocks are often found to be well in advance of the others. Provided that no disease is present in the apiary, a frame of sealed brood can be taken from the hive where room is needed, shaken clear of bees and exchanged with an empty comb from a weaker stock, or a frame fitted with foundation can be inserted on the outside of the brood nest. Add only one comb of brood at a time, otherwise the bees from the weaker stock may not be able to cover the sudden increase. The addition of the

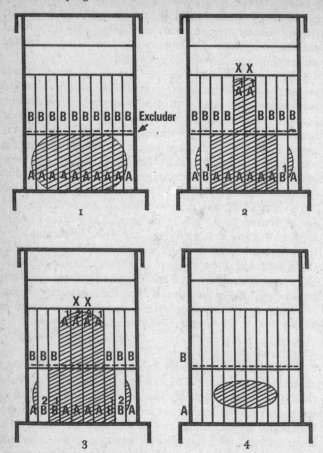

FIG. 25 Avoiding congesting with two brood boxes (shaded areas represent brood). 1 Put deep super above excluder. 2 First comb exchange. 3 Second comb exchange. 4 After flow box B a honey super. Only the combs marked X need inspection in upper box.

brood to the weaker stock assists its rapid development; the provision of extra room combined with removal of sealed brood from the stronger stock will tend to retard its progress towards swarming condition. .

The second alternative—use of a deep super above the excluder—allows greater flexibility and is a useful method for removing poor combs from the brood chamber. In the spring a second brood box, fitted with combs or foundation, is placed *over* the excluder and shallow supers are placed above that when necessary. Whenever room is required below, combs of sealed brood are shaken clear of bees and exchanged for frames of foundation or comb from the deep box above. Frames containing brood transferred to the upper box are inserted at the centre, but empty frames put into the brood nest are placed at the sides of the brood so as to maintain some approximation to the natural shape of the nest. Always shake or brush the combs you put above the excluder to avoid accidentally transferring the queen and drones. The operation can be repeated whenever necessary but the combs transferred above should always occupy the same central position. The brood in this deep super need not be inspected for cells, since certainly I have never had them built there. The procedure is illustrated in Fig. 25.

For those beekeepers who have chosen too small a brood chamber in the first place, this is one way in which they can enlarge the brood nest without the bother of examining twenty-two combs at every inspection. The system scores also by increasing the area on which young bees congregate. It must be stressed that this is not a swarm prevention system, but only a means of avoiding congestion. These extensions of the brood nest are seldom required with Langstroth and not at all with Modified Commercial, Jumbo or M-D. Now let us deal with the fourth purpose of our inspection.

Checking for queen cells

The regular checks for signs of swarming involve looking for queen cells on the faces, edges and bottom of combs. Systematic examination of every comb is unnecessary at every inspection until the day comes when cells are found. Routine inspection takes not more than three to five minutes per hive if you adopt the following routine.

1 Open up, putting roof, supers and excluders behind or beside the hive.
2 Lever the frames across the hive to one side, then push back the outside frames, nos. 1 and 2, which need not be taken out. Remove and examine frames nos. 3, 4, 5 and 6, successively. When you see queen cups, turn the frame to the vertical position and look inside.
3 If several combs show empty queen cups but no cells, there is no need to go further; stop examining when you have reached one comb past the centre of the nest and close up. Tighten up the frames, and replace the excluder and the supers.
4 If cells are seen to be occupied, then take positive action by one of the methods of swarm prevention, which should have been decided in advance.

Swarm prevention, general principles

Swarm prevention methods usually require the destruction either of all the queen cells or of all but a few chosen cells. In order to accomplish this, every comb of the hive must be cleared of bees and minutely inspected so that no cell is missed. If a single cell is overlooked, your plans may go sadly awry. Never, never destroy all the queen cells until you have made certain that there are eggs present in the brood nest. If you wish to keep some of the cells, use open cells and mark the comb or combs with a drawing pin in the top bar. These are basic principles for swarm control, not in themselves a method.

One other manipulation needs to be mastered—finding and isolating the queen. This has been explained previously. Clearly the disturbance created by cell destruction would prevent her from ever being found after shaking the bees from the combs, so that if the chosen system entails finding the queen, this must be done, using as little smoke as possible, before the cells are destroyed, and the comb on which the queen is laying must be put in a spare brood chamber, nucleus hive or box until the cell destruction is complete. With the mastery of these two techniques, finding the queen and finding every single queen cell, we can confidently face any of the multifarious systems of swarm control ever devised.

There are two fundamentally different types of control methods. The first brings about artificially the division of the colony into two or more units, in one of which we reproduce, as closely as possible, a swarm with the old queen, the flying bees and supers, while the other half of the stock becomes one or more nuclei of brood, young bees and a queen cell. These methods are based on making an artificial swarm, and they necessarily involve increase in the number of stocks, at least temporarily, and the employment of additional equipment. The second type includes those systems which frustrate the swarming impulse and allow re-queening during or after the main honey flow, using no extra equipment, or at most a cheap nucleus hive.

It is not the intention of the book to provide a complete guide to all swarm control methods. If the two main systems, fully explained here, are mastered, they can be used for the rest of a beekeeping career just as they are, or used as the basis for any other variant which the reader cares to experiment with, some of which will be outlined.

The artificial swarm

To carry out this manipulation you will need to have a spare floor, brood chamber, cover board and roof. As soon as queen cells are discovered, move the hive to one side, and in its stead set up the new floor and brood chamber complete with frames or foundation. Remove two of them from the new box. Now go through the stock and find the comb with the queen, carefully examine it without shaking or disturbing the bees, remove any queen cells, and place it in the centre of the new hive. The old queen is the hub round which the artificial swarm will be built. Now add a comb of honey and pollen without brood from the old stock, destroy any cells which may be on it and go through the combs in the parent stock; mark with a pin the top bar of the first comb you can find with an open cell containing a larva floating in royal jelly. Shake the bees from two other combs into the swarm but put the frames back into the original hive. Then rebuild the new hive with the 'swarm', adding excluder and supers, complete with their bees. Now finish going through the parent stock, shaking off the bees and destroying all the cells, except the marked one. Fill the box with the two spare frames removed from the 'swarm' before closing it up. The parent stock is now reduced to a single brood chamber with one queen cell, no queen and no supers. Carry it to a new stand a few metres away and facing in a different direction. As in the natural swarm, the foragers will fly back to their old position, where they join the swarm. In course of time, the new queen will emerge, mate and head a new colony. Those who are using the deep super can put the old queen into it instead of finding another brood chamber and, as it is likely to contain stores it will probably not be necessary to put in a comb of food, nor to shake in combs of young bees; just remove a comb, put in the comb with the queen and cover with the excluder and the other super. The steps for the artificial swarm can be summarised as follows:

1 Move old hive to one side.
2 Place new hive on old stand and add to it:—the old queen on her comb—the shaken bees from two combs—a comb of food—a full complement of frames with comb or foundation.
3 Put queen excluder and super back over the hive with the 'swarm'.
4 Destroy all except one open cell on one comb in the parent stock.
5 Move parent stock to a new position and direction.

An artificial swarm differs in two major respects from a natural one. First, it does not leave the hive and have to be caught; secondly, the division of bees into a random selection across the ages and stages of workers becomes an artificial division into young and old bees. For this reason the addition of the young bees shaken from two combs is needed to redress the balance in the swarm. In both natural and artificial swarms, the new queens, in their cells, spend their last few days in a much depleted colony.

It is possible to make an artificial swarm before the bees make swarm cells, but the practice is to be deprecated because the bees might have gone through the season without swarming. Furthermore, a queen entirely raised in the 'parent' stock, with only half the number of bees, is unlikely to be as good as one raised under the more prosperous conditions which lead directly to a swarm queen.

Systems based on the artificial swarm

The slight defects of the artificial swarm have led many beekeepers to try to remedy them by ingenious manipulations, all of which require very accurate timing in their operations. The reader is advised to refer to specialist works for full details.

Mr Armitt, when he found cells, delayed any further move until one day before the most advanced queen cell was capped. He made an artificial swarm and placed the parent stock on top of the supers on the old stand until one day before the new queens were due to emerge. He then moved the parent stock, as one or several nuclei, to a new stand. The system is designed to allow the bees to divide more naturally between the swarm and the nucleus, and to raise queens in a prosperous environment up to the last minute.

L. E. Snelgrove devised a most ingenious and elegant system with a special screen division board provided with six wedge entrances, arranged in three pairs above and below the board. He made an artificial swarm before he found queen cells, and placed the parent stock above the screen board and supers with one entrance open—to allow the foragers to fly and rejoin the swarm below. At weekly intervals the top wedge is replaced, the one under it opened, and a top wedge on another side of the hive opened. In this way bees are 'drained off' from the top brood chamber to the lower and the queen can eventually fly and mate from the third of the upper wedges opened. The system sounds intricate, but when mastered is quite simple and involves less opening of hives and disturbance to the bees than any other I know. In urban or suburban areas, when bees are kept at home, it is to be recommended, but entails too much travelling at short intervals for distant out-apiaries.

Swarm frustration systems

It will be noted that all the preceding methods based on the artificial swarm necessitate a doubling of the number of bee colonies during the summer months. The divided colonies can, of course, be united after the flow. They involve also the use of two brood bodies, roofs, floors and cover boards per colony, and this means a bigger capital investment than may have been contemplated. The multiplication of colonies means that a reduction in the total output must be expected. A colony that has been kept in one unit throughout the season yields more honey in one hive than the two separate divided units, and with less capital tied up.

Apiarists have tried for many decades to devise a method of frustrating the basic reproductive urge of the bee colony. Devices for trapping the queen as she left the hive have been tried and dropped. Various non-swarming hives have come and gone when the purchasers discovered that the hives indeed did not swarm, but the bees continued to do so. Others have tried clipping the queen's wings to prevent her flying and leading off a swarm. An attempt to prevent swarming by retaining the queen, without changing any other conditions in the hive, merely delays the swarm by the eight days between sealing the queen cell and the emergence of a virgin. As a method of swarm prevention it is useless, but in out-apiaries it is often advocated as a time-saver. We want to reduce our number of visits to the minimum to save time and fuel. If the queen's wings are clipped we can be quite sure that no swarm can leave the hive for at least fifteen days after we have ascertained that no cells are left. The clipped queen comes out with a swarm, perhaps eight or nine days after the inspection, and, unable to fly, falls to the ground and gets lost in the grass. The swarm circles, tries to cluster but fails to attract the queen, and returns to the parent colony to await the emergence of a virgin eight days later. At least we have not lost the swarm and we can keep the colony together by reducing the number of cells to one, at our next visit in the fifteen-day cycle. If we fail to return then, the swarm, headed by a virgin queen, may leave the hive. Hence it can readily be appreciated that reliance solely on wing-clipping for controlling swarms is bound to result in failure.

However, if we are prepared to use wing-clipping as part

of a system which includes standard seven to nine day inspections, then it will repay handsomely the little effort and skill required. We will examine the method point by point before setting out the practical directions.

Marking and clipping the queen

For many centuries man has docked the tails of sheep and dogs, trimmed the hair of horses—and his own—and clipped the wings of queen bees without ill effect on the subject involved. Carried out as a systematic part of swarm prevention, wing clipping is a simple enough procedure. I always make it a practice to clip and mark my queens at the first spring inspection. The queen is then easy to find; the bees less inclined to inhibit operations. When I go to the hives I carry round a small box containing a tiny tin of Humbrol paint, the kind used for model aeroplanes, a few match sticks and a pair of sharp needlework scissors. The paint tin is first stirred, and then reversed several times with the lid on—as soon as a queen is found, I take her up by the thorax, with the thumb and forefinger of my left hand. I then take up the scissors and cut about two thirds of her fore and rear wings on one side. The left side is cut when the queen was raised in an odd number year, and the right side in an even number year. Still holding the queen, I open the lid of the paint tin and pick up a small quantity of paint on the end of a match stick from the lid; I then rub this very small quantity of paint onto the top of the queen's thorax. There is an international agreement on the colours to use in any year and I always use the correct colour as it helps me to date the queens with accuracy. The last digit of the year determines the colour to use:

0 or 5	Blue
1 or 6	White
2 or 7	Yellow
3 or 8	Red
4 or 9	Green

As soon as the queen is marked, I blow on her just a little to harden off the paint and let her run down between the combs. I then mark up the record card: cl.m. The first step in swarm prevention is completed right at the beginning of the season.

Cell destruction

I now embark on the second stage of the system, namely routine seven to nine day inspections, checking as usual for food, eggs and general development, taking such steps as may be required for feeding, providing space, supering etc., that have already been described. As soon as I discover cells for the first time, I destroy all of them without exception, shaking each and every comb clear of bees to ensure a thorough job, as there can be no margin of error. The record card is then marked CX—cells destroyed; this is stage three.

On the next visit after CX, I start by looking for eggs or very young larvae before looking for queen cells. This is the key to the whole system, because when queen cells are destroyed the bees must be left in a position to requeen themselves from eggs or young larvae. Failure to leave them a potential queen will lead to the eventual extinction of the colony. Hopelessly queenless bees gather little or no honey. On the other hand, all existing cells may safely be destroyed so long as eggs are present.

Having ascertained that eggs are present, the second batch of cells can be destroyed, following the same procedure as before: shake the bees off every comb in turn, examine thoroughly for cells and once again mark CX on the record. The beginner and others who have never tried it may have visions of repeating this operation on every hive throughout the season. Fortunately, in practice the cell-destroying operation is seldom required more than two or three times for any one hive. Sometimes, because of changed conditions, the bees may give up the idea of swarming and not build a second batch of cells. The colony then returns to the previous stage, requiring normal seven to nine day inspections. At other times, after a second batch of eggs has been destroyed, the bees cease to feed the queen for egg laying and no eggs will be found in the worker cells. This is the sign the beekeeper has been waiting for. As soon as no eggs can be seen, stage four has been reached. Mark the first *open* cell, in which you can see a good larva and plenty of royal jelly. A drawing pin stuck into the top bar just above the cell marks it very clearly. Do not destroy any cells until the desired one has been chosen and marked; it is then necessary to go through the combs and remove all other cells, leaving only the one.

The final and fifth stage will be reached at the next inspection, seven to nine days later. First check that the marked cell, now capped and probably 'ripe', has come to no harm; then remove any emergency cells that may have been thrown up over young larvae. The hive can then be safely closed down and not opened again for several weeks, when it is wise to check that a new queen is laying. The record can be marked 'CCL Final'. The reader who has read the preceding chapters attentively will be able to work out precisely what happens in the seven to nine days between stages four and five: as soon as the marked cell is capped the old queen leaves with a swarm. Since her wings have been clipped, she is unable to fly more than a few feet and gets lost in the grass, the queenless swarm returns to the hive and no further swarm can issue until a virgin emerges. Hence the purpose of our last visit must be to remove all other cells except the one we have chosen. No swarm will leave when the virgin appears unless another queen cell is left behind.

A word of caution about stage five. It sometimes happens, though not often, that a clipped queen finds her way back into the hive and starts laying again. This puts the colony back into stage three—where you find cells and eggs and you need to remove all cells including the marked

one. The alternative is to find and remove the queen—
you can then leave an open cell as at stage four with a
further stage five visit later.

Many commercial beekeepers work to just such a
schedule as this, but include refinements of their own for
requeening or making nuclei. Many amateur beekeepers
and writers on beekeeping matters condemn the cell
destruction method of swarm control either because of the
number of inspections involved or because it leads to the
stocks becoming listless and queenless. This is not so. I
have stressed that the bees be allowed to retain eggs or one
queen cell. Under such conditions the colony remains a
fine vigorous honey-gathering unit throughout the season.
As for the work, it must be admitted that some is involved;
it always is with every form of livestock. But if one's interest, whether commercial, recreational or scientific, lies
with bees, it seems to me quite reasonable to expect to
devote about ten minutes per week to each stock during
the swarming season, which in England lasts from the
beginning of April to the end of July and is much shorter
in many countries.

The system has far more points in its favour than any
other. Among these we may note:

1 A colony which does not prepare to swarm is left alone
  to develop naturally.
2 No additional equipment is required.
3 Colonies which prepare to swarm are maintained as
  producing units until after the honey flow and are then
  left with a naturally raised, young, mated queen.
4 Bees are not forced to build cells against their natural
  inclination.
5 All the work involved can be fitted into the normal
  seven to nine day inspection rota without additional
  visits.
6 The number of hives requiring inspection diminishes
  as the season advances.

7 There is normally no need to find queens at swarming time, when they are notoriously difficult to find.

In summary the system involves these stages:

Stage 1 Clip the queen at the beginning of the season.
Stage 2 Carry out routine seven to nine day inspections.
Stage 3 When cells are first found, destroy all of them. Be sure eggs are present before destroying cells. Repeat as long as eggs are found on routine inspection.
Stage 4 When eggs are not found, leave and mark one open cell, destroy others.
Stage 5 Make a further visit seven to nine days later and remove all cells except the previously marked one.
Stage 6 Inspect a month later to see whether the new queen is laying.

Some may object that the system allows the old queen to die, when it may be their very best queen. This is true enough, but so does any other system of swarm control or even natural swarming. A queen from any colony which swarms or is artificially swarmed is doomed to die or be killed by the bees within a few weeks, and little can be done to prevent it. If for some special reason you feel you must preserve an old queen, then she can be removed with the frame she is on and used to make a nucleus at stage four, when one open cell is left in the hive, but this should be the exception rather than the rule. In natural bee life, humanitarian principles have no place and it is better to let nature take its course, under proper control. You have, in any case, the advantage of a new queen, daughter of the one you have lost.

On several occasions when I have advised that an open cell should be left, it could be argued that a capped cell would advance the colony by several days towards its eventual requeening. Unfortunately, this is not always the case. Queens do die in the cell, or an evacuated cell is sometimes resealed with a dead worker inside. The capping hides the contents, and attempts to remove the cover result in damage to any larva or pupa which might be inside. At the time when swarming occurs, a few days delay in re-queening tends to be beneficial rather than harmful, and

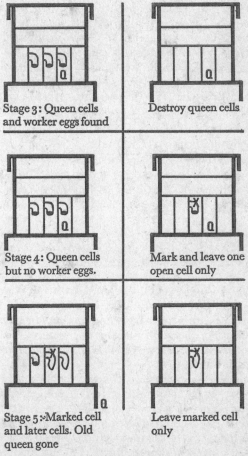

Stage 3: Queen cells and worker eggs found

Destroy queen cells

Stage 4: Queen cells but no worker eggs.

Mark and leave one open cell only

Stage 5: Marked cell and later cells. Old queen gone

Leave marked cell only

FIG. 26   Swarm control without increase

you can look into the open cell to be sure that it contains a good healthy larva with plenty of royal jelly.

Mating the queen

When a colony contains a virgin queen, the utmost care must be exercised to see that she can be mated without let or hindrance. She may make one or more successive reconnaissance flights to locate the hive entrance. On no account move the hive, or change any features around the hive, until she has started laying. This means that the hive itself is not moved in any way and that other hives, salient plants and garden features have to remain in situ. It is advisable for the beekeeper to keep away from the hives between 10 a.m. and 3 p.m. The presence of a human shape in the vicinity of the hive upsets the picture of the location imprinted on the queen and sometimes deceives her into rejoining the wrong hive, leading to her immediate execution.

Queenless colonies

A normal colony, in spring and summer, usually replaces a failing queen either by swarming or by supersedure, but occasions nevertheless arise when the absence of eggs and brood lead to the assumption that the colony has become queenless. Every year, in the autumn, I receive a number of calls from distraught beekeepers who assure me that their colonies are queenless and could I please spare them a queen. The end of the honey flow signals a sharp decrease and, not infrequently, a complete cessation of laying. The queen is probably present, the colony healthy and there is no need for alarm, but it is wise to make a very simple check for queenlessness by inserting into the centre of the hive a comb containing eggs and larvae from another stock. If the colony is, in fact, queenless, the bees will quickly raise emergency cells on the young larvae.

Inspect the colony four or five days later and remove any sealed cells, leaving one open cell for the bees to requeen themselves; or they can be requeened with a new queen from some other source. This simple test is applied at any time that queenlessness is suspected. If no cells are built the colony can be assumed to be queenright. No requeening should be attempted until the old queen has been destroyed. A colony with laying workers cannot be requeened, but the bees can be shaken onto a running-board in front of another hive.

Requeening

The colony consists of a queen and her daughters, and any disturbance of that arrangement causes an imbalance which continues until the new queen is once again surrounded by her daughters. If a new queen is introdced, she is incontinently put to death. Even when she has been apparently accepted, a strange queen lives only on suffrance for about six weeks after the introduction. For this reason, queen introduction is only resorted to in order to change the breed, to increase the number of colonies by forming nuclei, or as part of a swarm control system. I would like to stress that requeening fails most frequently after the end of the swarming season and before the bees settle down in the autumn with their winter stores. In Britain the dangerous period runs from the beginning of July to the end of August and beginners should endeavour to requeen nothing bigger than a nucleus and to requeen their stocks with nuclei instead of just queens at that time of year. Without going into a full description of all the methods that have been devised for requeening, a few principles need to be mastered before starting this difficult operation. The stock must be made queenless, either by removing the old queen and killing her or by making a nucleus with her. The new queen is best introduced between one and two hours after the stock or nucleus has been left queenless so that the bees have had time to become aware of their queenlessness and anxious to receive a new queen. Finally, the new queen is introduced in a wire cage which allows the workers to feed the queen through the mesh and from which they can release her after a few hours when she has acquired the colony odour.

The simplest cage for the purpose can be made by using a hair roller or a piece of wire cloth, of three meshes to the centimetre, rolled into a tube about 16 mm diameter and 75 mm long. One end is blocked by a cork, and a small square of newspaper is fixed over the other end with an elastic band. The queen is placed in the cage and the paper held in place while the elastic is put round it. The whole cage is then lightly wedged between the top bars of two frames and the hive closed and left. The cage is removed from two days to a week later, when the paper has been eaten away and the queen released.

When queens are purchased, they arrive in small three-cavity boxes with twelve to twenty workers, which should not be introduced for fear that the strangers might arouse aggression which could be turned against the new queen. We solve the puzzle of removing the queen only, without her flying off, by resorting to a simple stratagem. Get into a car, close all windows, and put a tea-towel or cloth over the heater vents above the dash board. Open the postal cage by removing the gauze on one side. The queen is then picked up as she leaves the cage; but if she escapes, she will fly towards the light and can be caught on the windscreen and put into the introducing cage. The same result is achieved in a shed or room with all the windows and doors shut. A word of warning: some window cleaning fluids contain DDT or other insecticide which could be fatal to a queen who touched it. Work near the light and the queen will always fly to the window. If the colony is first dequeened, the transfer of the queen from the postal cage to the introducing cage can be undertaken at leisure, leaving plenty of time to insert the cage at the optimum time.

However, introducing a strange queen to a full stock is at best a risky enterprise. It is easier to introduce a queen to a nucleus and, if desired, introduce the nucleus to the full colony at a later stage when she has started laying and is surrounded by her own daughters who will protect her.

Making a nucleus

A nucleus, as we saw in a previous chapter, consists of three or four combs of bees, food and brood with a laying queen. We have to know how to make nuclei with unfailing success. We need a nucleus hive or a spare hive divided by a wooden dummy to restrict the space to the number of combs used. We go to the hive we wish to divide, find the queen and confine her temporarily in a queen cage, or put the comb she is on into another spare hive or box. Next we carefully take out a comb of honey and all its adhering bees, then a comb with a small patch of pollen and brood, and a third comb with sealed brood covering not more than a third of the comb area. Full combs of brood will result in the chilling and death of the pupae at the corners and ends of the comb. Shake another comb very lightly to dislodge the older bees back into their original hive and shake the remaining younger bees into the nucleus. Close up the large hive after replacing the missing frames and the queen. Block the entrance of the nucleus with green grass or herbage and put it on a new stand in the shade. After about an hour, open it up gently, using smoke, wedge the introducing cage between the centre top bars and close up again, putting a feed of thin syrup over the top. The grass wilts and is removed by the bees in due course. Some of the older bees will inevitably return to their old hive, but enough young ones will be left to start off the new colony. Because of this loss, many experts prefer to make nuclei and carry them to an out apiary where they can be opened immediately.

Nuclei can be made with introduced laying queens,

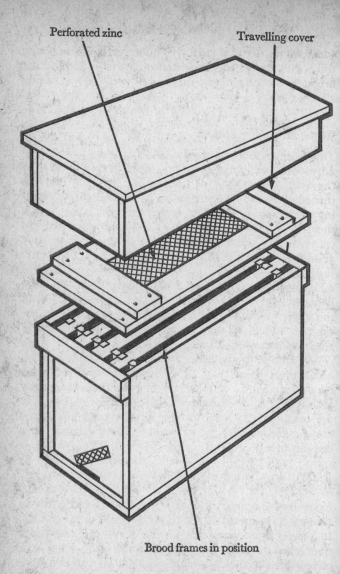

Perforated zinc

Travelling cover

Brood frames in position

FIG. 27 Nucleus hive

with virgin queens or with ripe queen cells. In every case the nuclei are left until the new queen is laying and her brood is emerging, before they can be used for requeening a large stock.

Requeening with a nucleus

This is always more satisfactory and safer than introducing a queen direct. You can unite a nucleus to a colony, by the newspaper method, at any time except the winter. Go through the stock and remove the unwanted queen, crushing her thorax to kill her. Cover the brood box with a sheet of newspaper, pierced with a few jabs from the corner of your hive tool, and place an empty brood box above the paper. Transfer the nucleus to this box, bumping in any bees which may remain in it. Place dummies on each side of the nucleus. In two or three days, when the paper has been chewed away, remove empty combs from the lower box and put the nucleus frames down in the centre as a unit. If there are no empty combs, any spares removed can be distributed to other colonies.

At times when the majority of foragers are out of the hive, which means in good weather during a minor or major flow, but never in bad weather or at the cessation of a flow, the introduction can be speeded up by removing the unwanted queen and by putting all the nucleus frames as a single unit into the lower box, about an hour after dequeening.

Queen breeding

Every beekeeper should rear a few queens every year to safeguard against failure or accident to one of the queens. It is even more important for the very small beekeeper than for the larger enterprise to insure against loss of a queen. This section is intended only as a guide for those who wish to raise a few queens from natural cells. Others who wish to pursue the subject will find a number of specialist books to guide them.

Natural cells, which are most frequently swarm cells, possess one great disadvantage. They may perpetuate the tendency to swarm. However, they do have an advantage, in the ease with which they are obtained and the excellence and vigour of the queens which emerge from them. With good records from previous years and careful choosing of the breeder queens, some improvement of the strain can be achieved. I will endeavour to lay down the general principles involved before summarising the practical steps.

A queen-rearing programme can only be initiated when drones have been observed in the apiary, although the natural cells are unlikely to appear until the bees themselves are satisfied that drones are around. Good queens are raised from strong stocks and at a time when stores are coming in. These are important points to bear in mind. Whenever possible, rear queens from a stock which did not swarm the previous year and which gave an above average yield of honey for the apiary in which it was situated. The beekeeper must be quite familiar with the timetable of queen rearing, from the laying of the egg to the emergence of the queen, and it will help if the appearance of the larva on each of its five days of growth becomes reasonably familiar—queens develop in the following way:

Days 1–3	Dry egg in cup.
Days 4–8	Larva on royal jelly, cell lengthens.
Day 8	Cell capped over.
Day 9	Swarm leaves with old queen unless prevented.
Days 10–15	Quiet period.
Day 16	Virgin queen emerges and can either fight and kill rival queens or leave with cast or delayed swarm.

Between days 19–30 Virgin leaves on mating flight and settles down to lay.

Virgins unmated after two weeks seldom prove to be of much value.

Our operations have to conform with this timetable regardless of the weather. We may have to cover the hive with an umbrella or surround it with a windshield, but open it we must for at least two of the major operations involved.

Queen rearing starts as part of normal swarm control by clipping the queen's wings in early spring and then inspecting the colony at seven- to nine-day intervals for queen cells. As soon as we find queen cells in the selected colony we make an artificial swarm, but without the need to shake any young bees over the queen. The stock is then rebuilt with the new swarm below, then an excluder and the super or supers. If possible a second excluder goes over the supers and the brood placed over that excluder. Before closing the hive the brood needs to be rearranged in this upper box so as to make one, two, three, four or even five nuclei according to the number of queens required. The minimum number of combs must be two for each nucleus, but three is a better, more viable unit. Each nucleus group in the upper brood box requires an open queen cell; brood; pollen; honey.

When examining the combs, brush the bees off rather than shake them, which might damage the queen larvae. If any capped cells are found they must, of course, be destroyed. No division boards are required—we merely note in our record 'Divi $3 \times 3 + 2$', for instance, which indicates that the top brood box has been divided into three three-frame nuclei with two combs left over.

Assuming that the oldest larvae are about to be capped next day, we then have an absolute maximum of nine days before the first virgin will emerge. Before that happens, the nuclei must be taken off, placed into separate nucleus

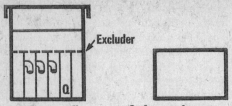

1 Queen cells appear – find empty box

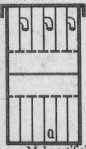

2 Make artificial swarm – arrange
series of nukes above the supers,
each with an open cell

3 One week later, remove nukes to
nucleus hives

FIG. 28 Simple queen rearing

hives and removed to different stands, where the virgins
can emerge and be mated. The original stock will behave
as a normal artificially swarmed stock and will probably
supersede the old queen a few weeks later.

Remove the mating nuclei, or 'nukes' in beekeeping jargon, to another apiary if possible or to a friend's garden 2 km away for a few weeks, until the new queens are laying. If there is no chance of finding a distant locality, plug the entrances of the nuclei with green grass or soft green herbage and face them into a different point of the compass. Give each 'nuke' a feeder of dilute syrup and check a week later that the queens have emerged. Renew the syrup whenever the feeder is empty, because small nuclei can ill afford many foragers out on distant journeys. Mating nukes should be looked at before 10 a.m. or after 4 p.m. During the midday period the new queens may be out on reconnaissance or mating flights, and their return to the hive may be disturbed, with fatal consequences, if the beekeeper is at their hive.

The uses of nuclei

A few nuclei in reserve can radically alter one's outlook on beekeeping for the better. There is no longer any undue need to fear accidental damage to a queen. A colony that fails to develop, a queen that begins to fail through old age—these can be remedied as soon as noticed by re-queening with a nucleus, usually in good time for the honey flow. At the end of the season two nuclei can be amalgamated to increase the number of stocks put down to overwinter. As experience increases, the way in which nuclei can be used will also be found to increase.

Obtaining early swarm cells

The small-scale beekeeper has perforce to wait until his bees start to swarm of their own accord, but if early queens are needed and one can afford two stocks of bees to be directed to queen rearing, cells can be induced almost as soon as the first drones take the air.

Two strong colonies are selected, preferably close together or on the same stand. Make a three-frame nucleus with the least desirable queen and move it some distance away to develop as best it may. Move the queen-right stock to the space between the two hives, remove its supers and queen excluder, then unite the two brood boxes with paper, leaving on the supers of the upper colony. Put a second sheet of newspaper over the supers and put on the supers from the queen-right stock. Next day, when the bees have united, the supers can be reduced to no more than two by removing empty combs and boxes.

You then have two colonies full of brood and bees of two queens, but only one queen to cope with supplying queen substance and eggs for a suddenly enlarged nest; the reduction of supers also helps towards forcing congestion. At the next inspection, seven to nine days later, cells will be found in plenty and nearly twenty combs of brood for making nuclei.

An artificial swarm can then be made, using the queen from the combined stock to head the 'swarm' in the lowest box, while the two boxes of brood, divided into nuclei, can provide quite a number of queens.

If the double brood box holds more combs than are required these can be used up, either to strengthen the nuclei when these are divided off the following week, or to reconstitute a nest for the queen which had been removed to a new stand at the start of operations.

It must be clearly understood that queens raised in this way are reared at the cost of a considerable portion of the honey crop. The two strong stocks started with will be reduced to struggling colonies which may or may not produce a surplus on the main honey flow. If the new queens mate early enough, a nucleus can be added to bring them up to strength for a late flow. It costs less to wait for cells in their good time.

Beekeepers with several apiaries often make a practice of ensuring a succession of early queens by 'taking out a nuke' from colonies with the first swarm cells they meet in the spring. Instead of destroying all the cells, a comb with an open cell is placed in a nucleus hive, and two further combs containing pollen, honey and bees are added to the nuke, either from the same hive or from other hives which can spare them without becoming unduly retarded. The nuke is then taken to the next out-apiary and left there for mating. New nukes are taken out at each apiary for carrying on to the next. In this way each apiary acquires a number of nukes ready for any emergency which might arise. Queens produced in this way may not be the best, but they are always there in case of emergency, and can be replaced at leisure later in the year.

It has now become clear that bees raise queen cells under three different impulses:

1. *Swarm cells* arise from the bees' natural, seasonal drive to increase. These always produce fine queens, but should only be chosen from a stock selected for its desirable qualities.
2. *Supersedure cells*, built naturally as the bees' out of season method of replacing a failing queen. They are few in number and are constructed in the centre of a comb independently of the other cells. Supersedure queens are as good as their mothers.
3. *Emergency cells*. Bees accidentally deprived of their queen build queen cells based on ordinary worker cells containing very young larvae. They float the larvae out of the original cell on a bed of royal jelly until it occupies the new vertical extension. In strong stocks, such cells produce perfectly good queens, in my experience, but small or weak stocks seldom produce good emergency queens.

I am aware that the manipulations and practices described in this chapter cannot easily be absorbed just by reading, but when the reading has been translated into practice the systems will fall into perspective and the difficulties will be found to lie more in the apprehension than in the execution.

8

Honey and Wax

Bees gather nectar during the flowering season and convert it into honey which they store against the months when no flowers bloom. In temperate climates this means the winter, but in arid, hot climates the reserves may well be consumed during the hot summer.

If the beekeeper is to share some part of the sweets, he must earn his portion by managing the bees in such a way that they produce more honey than if they were left to their own devices, and more than they need for their own welfare. We call our share of honey 'the surplus'; surplus, that is, to the needs of the colony that gathered it. In practice the brood chamber, when it is really full, holds enough for the bees, and the honey stored above the excluder can be looked upon as our reward for inducing the bees to increase their production. Any beekeeper who manages to obtain a large surplus from his bees as well as leaving them fully provided for the winter can take pride in his achievement, and those who tell you, as they surely will, that they just leave their bees to 'get on with it', deserve the poor results they get.

When to take the honey

The honey crop is taken once a year, a week or two after the main honey flow, and no one can expect to keep his bees in good heart by constantly removing combs of half-sealed honey during the course of the season. Unsealed honey should never be looked upon as stored honey and seldom remains stationary in the combs. Bees leave cells open either because the honey is not ripe and will not keep or because they require it for food for immediate consumption. We may have placed a queen excluder over the brood chamber to divide off 'ours' from 'theirs', but so far as the bees are concerned the nest is one continuous unit with the brood at the centre, surrounded by an inner ring of pollen and an outer circle of honey extending as far as the top of the highest super, to be drawn on as need arises. In districts where the weather is uncertain, stores in the supers play an important part in the well-being of the bee community and may decide to some extent the bees' ability or willingness to go out and gather more. The more they have, the more able and willing they appear to be to add to the store. It would be unwise, therefore, to remove stores until the honey flow ends and the bees have sealed over all but a few cells in the storage combs and show that they know that the season has ended. The autumn flowers, michaelmas daisies, ivy and others, will carry them through to the winter. However, there are some exceptions to this rule, arising from the nature of the honey, which oblige us to remove it somewhat earlier.

The origin of honey

Honey is a complex end-product which starts with the sugars exuded from floral nectaries and gathered, elaborated, transformed and concentrated by the bees. The 'honey plants' have evolved over millions of years side by side with bees to produce a refined interdependent life system. Plants retain vigour in the strain by cross-fertilisation between different plants of the same species. The pollen from one flower has to be transferred to the stigma of a flower of the same species on another plant, and bees have evolved to play the part of long-distance pollen carriers.

Flowers attract bees by producing the sugars needed by the adult insects and a superabundance of pollen, enough both for pollination and for the bees to take home and use for rearing their brood. But for this process to be successful the bee must visit flowers of one species only, in any one journey. Some wild bees meet this requirement by concentrating their attention on a single species of flower during the whole of their lives. Others, like our honeybees, have developed a 'flower constancy', and visit only one type of flower so long as it continues to yield nectar or pollen in abundance; but as soon as the supply ceases, they can switch their loyalty to another, more productive flower. This constancy has most important economic consequences and turns our honeybees into valuable pollinators for some agricultural crops.

The weight of nectar a bee can carry amounts to some 30 to 40 milligrams per journey, and this will be gathered from one to a thousand different blooms according to the species. Of the nectar gathered, from one quarter to three quarters consists of water which is later evaporated, so that something like 40 000 to 50 000 bee journeys are required to add 1 kilogram of honey to the stores in the hive. If we go on further to calculate that a colony of bees requires 75 to 100 kg of honey for its own use and may yield a surplus of 40 to 50 kg to the beekeeper, we arrive at the staggering total of some seven to ten million bee journeys each season for honey alone, and perhaps half as many again for the pollen. A weak stock of bees is unable to achieve these results; it just does not have the necessary 'bee power'. Moreover, a colony which reaches its peak after the main flow may have the necessary numbers, but the locality has ceased to yield the nectar. I hope these figures may have brought home to the reader how necessary it is to bring bees to their peak foraging force early enough to take full advantage of the local honey flows, just as the previous chapter emphasised the need to maintain the foraging force in a single unit.

Composition of honey

To ensure that the consumer should buy only first-class honeys the Food and Agriculture Organisation, European Committee, have provisionally established a scientific basis for evaluating the degree to which honey has been heated, and the trace elements detroyed. One of the enzymes concerned, diastase, diminishes rapidly with heating and the substance hydroxy-methyl-furfurol increases. The FAO have laid down a ratio of these substances for various honeys, beyond which they are considered to be saleable for cooking or baking purposes only. Whilst the intention of the standard is to be highly commended, the basis must be questioned, since ageing without heating produces similar results and many quite natural and satisfactory honeys would fail the test. This diastate/HMF ratio concerns the commercial honey producer more than the beekeeper producing for his own and friends' consumption, but at least it shows us that once the honey has been produced it is only too easy to spoil it by overheating during subsequent treatment. Natural, untreated honey is a nutritious, healthy food possessing valuable medicinal properties as well as unequalled subtleties of flavour for the palate of the initiated gourmet.

Varieties of honey

Liquid honeys are generally grouped into three colour grades, light, medium and dark, with a separate class for ling honey.

The lighter grades usually derive from the clovers, limes (basswood), acacias, brassicas, willow herbs, sage, tupelo and some of the eucalypts. Their flavour tends to be delicate and subtle, and they command the highest prices because of their eye-appeal. The consuming public obviously prefers them to all others. Preference on the basis of colour is a poor criterion for a food that should be judged on its flavour, and may account for so many people asserting that they do not like honey because it is too sweet and tasteless. Light honeys frequently granulate with a very fine grain.

The medium honeys include a very large number of flavours and plant sources, but amongst others we can name most of the tree honeys—maple, sycamore, apples, plums, cherries, soft fruit—as well as a vast number of composite flowers, from dandelions to michaelmas daisies. They tend to be somewhat thicker and to granulate more coarsely than the lighter types, and have a richer flavour.

The darker honeys come from chestnut, blackberries, buckwheat and a number of 'bushy' plants throughout the world. The flavour is often quite strong and the granulation more frequently coarse than fine. The honeys we gather are seldom one-source products, and variation in colour and flavour can most frequently be accounted for by admixture of different honeys in the comb.

One that stands out from all the others is the honey produced by ling in the European moorlands and commonly called heather honey. It has the peculiar quality of thixotropy, or ability to gel when undisturbed and to run as soon as it is stirred. As it does not run, it cannot be extracted in the normal way but has to be squeezed out of the comb under heavy pressure, enclosing air bubbles which become trapped in the gelling honey and remain in it. Its thixotropic nature arises from proteins in the honey. It certainly has the most powerful aroma of all and a trace of bitterness which counters its sweetness and endears it to some palates as much as it repels others.

Honey-dew

Although bees gather most of their stores from floral nectaries, there are times when this supply fails and they may have to manage with natural sugars from other sources. A number of plants exude nectar from glands situated in areas other than the flower. The cherry laurel, often used as a hedge shrub, bears four distinct brownish patches situated under the leaf, on either side of the midrib towards the base. Field beans and broad beans possess a spear-shaped bract, or stipule, at the base of the leaf and flower stalks, with a central brown or purple nectary on its underside. Both these plants produce abundant leaf nectar prior to flowering, and bees gather the sugars all the more avidly because they both occur during a gap between the spring and summer flows. In some areas, moorland bracken produces quantities of very dilute nectar from glands on the leaf axils, when the fronds are young, and it occasionally happens, during warm springs, that it can contain enough sugars to attract bees. When processed by the bees and stored, all extra-floral honeys are known as honey-dew.

The most usual source of honey-dew has quite a different origin. Most of us are familiar with the aphids or plant lice which weaken and destroy many of our garden plants—the greenfly, the black fly and others. Many closely related aphids feed without causing much damage to their hosts on many of our trees—oaks, linden, beech, conifers and hickory all support large populations of such insects, which feed by inserting their beaks into the plant tissues. The pressure of the plant juices, which contain natural sugars, literally drives the food into the aphid's mouth until it cuts off the supply. It takes from the plant whatever nutrients it requires for itself and elaborates in its system a number of new sugars, over and above its own needs. The surplus is then eliminated in droplets which fall as a sticky liquid to the surface of the leaves, where bees may gather it.

Some honey-dews are greatly prized, particularly those gathered in the pine forests of Central Europe, where the product known as 'Tannenhonig' commands the highest prices. Other honey-dews are often less appreciated, mainly because of their frequent dark colouring, derived from soots, dusts and fungus spores which adhere to them while they lie exposed to the air on the surface of the leaves. Honey-dew produced from the lime trees bordering our city streets can be quite objectionable on account of the large content of industrial smoke particles, but honey-dews collected in the open countryside can seldom be distinguished from good-quality medium honey. It is quite probable that many more honey-dews are gathered than has been realised in the past. The sugars produced by the plant, elaborated by the aphid and later by the honeybee, have all the plant flavours and most of the essential oils and attributes of floral honeys, provided that these have not been overcome by contaminating substances.

Crystallisation

All the honeys are gathered, elaborated and stored in a liquid state, but after extraction they crystallise or granulate within a time which depends on the proportion of the different sugars contained in them and on the ambient temperature. As granulation is usually the least understood part of honey handling it would be well to explain it more fully.

Of the three principal sugars contained in honey, glucose is the most prone to crystallise. The greater the proportion of glucose, the quicker and harder the honey sets, and low temperatures favour rapid granulation. Furthermore, the faster the honey sets, the finer the grain of the crystals and conversely, the slower it sets, the coarser the grain. So those honeys that contain a high proportion of glucose will set rapidly, producing the fine grain that is so highly appreciated. Granulation is usually triggered off by nuclei, which may be specks of dust, wax or pollen grains but are more often tiny crystals. We shall make use of this knowledge when preparing to bottle granulated honey. A jar of naturally granulated honey soon shows signs of streaks along the side of the jar which are known as frosting. This makes the honey look unsightly and is often mis-interpreted as fermentation, but in fact frosting occurs as a result of temperature fluctuations which cause the honey to contract and pull away from the jar, allowing air to fill the gap. The honey oxydises in contact with the air, bleaching it and bringing about the frosting streaks but in no way impairing the quality of the honey. The same oxidation always occurs on the top layer of a jar of crystallised honey, but as the colour change is uniform it is not viewed in comparison with the unoxidised part and therefore passes unnoticed.

Brassica honeys

Most of the brassica or cabbage family yield honeys with such a high proportion of glucose that they granulate very quickly indeed. With the increasing culture of oil-seed rape for milling into salad and cooking oil, the possibility of moving bees to pollinate and gather a honey crop from these flowers has improved the outlook for many bee-keepers. It has brought many problems too, in exposing bees to the insecticide sprays used on the crop and in dealing with a honey which granulates in the comb even before it has been extracted. In the case of rape honey, an exception must be made to the rule of waiting until the end of the year for removing the honey: it has to be removed and extracted immediately the bulk of the honey is sealed.

Removing honey from the hive

There are a number of ways of clearing bees from the supers. Carbolic acid cloths were used in the past, and recently propionic anhydride and benzaldehyde soaked cloths have been used. Benzaldehyde succeeds under some weather conditions but not others, and I have come to the conclusion that no method of clearing supers is so neat and efficient as the Porter-escape clearer-board. A single one bee-way escape may get blocked, but two two-way escapes have never failed, in my experience, to clear the bees from all the supers overnight. The supers are lifted off and the clearer boards placed over the queen excluder, making sure that the escapes are well bedded down in their slots. If they are at all loose, they can be fixed down with a drawing pin at either end. The supers are then piled back on top. If New Standard or similar ventilated roofs and covers are used, a piece of paper should be placed over the cover board to prevent bees trying to move out towards the light coming in from the top. If you have insufficient clearer boards for all your hives, no harm will come from piling the supers from two hives on to one brood chamber. The bees seldom seem to fight and on their next flight out they will return to their original home.

Naturally, the alignment of the supers is checked to prevent any gaps through which bees might rob out the honey. At any time the following day the supers are prised apart and carried to the transport or store. It is advisable to carry a number of cloths or boards to cover up the supers. When carrying supers hold them vertically, top bars against the body; the weight distribution in this position allows them to be carried with less effort than would be required if they were held horizontally. The honey is sealed and the frames propolised together: there is absolutely no danger of spillage. The supers are taken home and stacked in a bee proof pile awaiting extraction. When the supers are off each hive, remove the clearer board and excluder, and put back the cover board and roof. Collect up the Porter escapes and keep them separately in an airtight tin.

The extractor

The extractor is possibly the most expensive single piece of equipment the beekeeper will ever use, and some

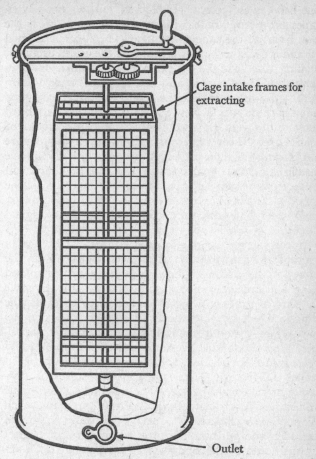

Cage intake frames for extracting

Outlet

FIG. 29 Simple honey extractor

thought is needed before making the purchase. Essentially it consists of a tall can with a geared revolving cage, deep enough to take the vertical length of the frames and to allow the honey to accumulate below the cage before running it off into the storage tins. The extractors normally supplied will accommodate the frames of National and Smith hives but have insufficient clearance under the cross-bar for Commercial, Langstroth and Jumbo frames. For these it is essential to purchase a Langstroth size extractor which can also be used for the smaller British frames. They may be hand driven or motorised, and tangential or radial. The choice of motive power will depend on the size of the beekeeping activities contemplated. If there are less than twenty hives, the labour of extracting by hand can be contemplated without misgivings. but if forty or more hives is the ultimate goal, then the motorised unit may justify its expense. In general, hand-operated machines should be of the tangential type, which require few turns for extracting the honey but extract from only one side at a time. The radial extractor better suits a power drive, because the longer extracting time, both sides together, involves no additional effort once the machine has been loaded. Different sizes of machines are available, but the most useful of all for general purposes is the kind known as the tangential 3/6, holding three deep or six shallow frames. The small beekeeper will manage quite well with a 2/4, but the 3/6 is more flexible and universally acceptable as a hand extractor. The motorised radials vary in capacity from 12 to 100. But to keep the larger machines gainfully occupied an uncapping machine is essential.

Whichever extractor is used it should be mounted and bolted to a strong bench or frame to raise it high enough to put a 12-kg honey tin under the tap. If it is not firmly mounted it will become unmanageable at times, while it is in operation. Of course, before the machine is used it has to be washed with water and detergent, thoroughly rinsed and left to dry. At the end of the extracting season it should again be washed and dried, and wiped over completely with a cloth soaked in medicinal liquid paraffin, and then enveloped in a plastic sheet or bag and kept in a dry place for the following season.

The uncapping knife and tray

Before honey can be centrifuged out of the comb, the cappings have to be shaved off with a sharp knife. Uncapping knives of various shapes and sizes, each with its loyal adherents, have come on the market. A sharp, serrated or scalloped, broad-blade knife is most frequently used by the small beekeeper. An uncapping fork also performs well, but the prongs break too easily for my liking. The viscosity of the honey prevents easy work with a cold knife. Even with a rapid sawing movement, the honey on the underside of the blade soon tends to break down the cells below and damage the precious combs. A pair of knives in a tall tin of hot water on a small stove or heated uncapping tray always gives satisfaction, but knock the water off the knife before using it by rapping the back sharply on the edge of the tin. Electrically heated knives, with an adjustable thermostat, are more expensive but perform better than any other makeshift arrangement.

The cappings fall from the knife in a thin sheet. They may still contain a fair proportion of honey, however near the capping we endeavour to cut. For the small beekeeper a very large plastic bucket or bin, fitted with a wire screen or sieve about 10 cm above the bottom, will hold quite a lot of cappings and allow the honey to drain off. Across the top of the bucket place a wooden bar, slotted underneath, to fit on the edges of the container without slipping, and with a cavity or spike on top to prevent the lug from sliding. The owner of a large apiary will invest in a more sophisticated Pratley uncapping tray. This useful apparatus has a heated water tray below a sloping false base. As the cappings fall on the hot metal they melt and fall into a receiving bucket where wax and honey separate, the wax floating on top and the honey falling to the bottom. Next morning the cake of hard wax is lifted clear and the honey is added to the bulk of the extracted honey for cleaning, or, if preferred, it can be set aside for cooking or mead making.

The extracting process

Extracting fills the air with the delicious perfume of honey. Bees and wasps from far and wide locate the extracting room with unerring accuracy within minutes of the start; it must be made bee-proof at any cost. If the work must be done in the kitchen, then wait until evening before uncapping a single comb. Warm honey extracts more easily than cold, and if a temperature of 25°C can be maintained, so much the better.

Take up a super, invert it with a bump on the work bench or table and lift it away from the combs. The frames are separated with a hive tool and uncapped. Cut as far as possible just under the capping, through the air gap. In the past, uncapping knives have always been made to cut upwards, a slight tilt of the frame allowing the cappings to fall away from the comb. This is entirely unnecessary and dangerous. The wax falls away just as easily from a downward stroke and there is no danger of a slipping knife damaging the operator's face. It is simpler by far to lean the comb away and slice downwards, using the 27-mm wide top- and bottom-bars as knife guides. Clear any recesses in the comb face with the point of the knife. When both sides have been uncapped, place the comb in the extractor and check that the honey valve is closed. Then continue adding combs until the cages are fully loaded. Load the machine evenly so that heavy combs are opposite heavy combs and light opposite light, otherwise vibration can cause damage to the brackets which bolt the extractor to the stand. When the load is complete, turn the handle slowly until the honey begins to spatter onto the sides of the barrel. A few more turns will empty about half the honey from one side. Then turn the combs round and empty the second side completely, finishing up with a few fairly brisk winds. Turn these frames back to the first side and finish extraction, winding fast enough to throw out all the remaining honey. The frames can be left to drip until they are replaced one by one with a new batch of full combs. Put the empty ones back into the super as you proceed. You will gather from this description that a plentiful supply of newspaper on the floor will be a great help at clearing-up time.

Late in the evening these supers can be replaced on the hives above the cover board and, if it is at all possible, they should go back on the hives they came from. Make sure that the hive is bee-proof, and endeavour to cause as little disturbance as possible while putting on the supers. They are placed over the cover board with the feed hole open. The bees will clear up the wet comb, storing the honey below the cover board and repairing the edges of the comb. The supers can stay there for the winter or be brought under cover.

Straining honey

The honey collects at the bottom of the extractor and is drawn off into a bucket or honey tin, together with bits of wax and debris which require straining out before the honey is stored. Put a 12-kg tin under the tap and drain off the honey. Honey runs silently and overflows without a sound, so run off a full tin and always close the tap before turning away to do any other work. The honey can be strained slowly cold, or more rapidly if it is warmed overnight in a tea chest or box with a shielded 100 watt bulb. The chest should hold four 12-kg tins.

So-called ripening tanks, with a coarse and fine strainer are often advised for straining and bottling. For the bee-keeper with only one or two hives these ripeners are adequate, but a better arrangement consists of a large plastic bin with a honey tap fixed to its base. A straining cloth of washed and wrung out scrim or nylon is tied tightly over the wide mouth with a piece of string long enough to go

two or three times round before knotting. Make sure the
scrim bulges down enough to hold a tin of honey in its de-
pression, then pour in a tin of warmed honey. At about
35°C the viscosity of honey breaks down and it runs very
freely. In a few minutes another tin can be added. If it is
intended to bottle the honey at once, leave it to stand for
24 hours to allow the bubbles to rise to the surface.

Bottling honey

Honey keeps so much better in large lever-lid cans that it
is wise to bottle only a month or two ahead of consumption and to store the rest, strained, in tins. Bottled honey
should be looked upon as a perishable commodity. True,
its quality does not deteriorate, but its appearance does.
Liquid honey begins to crystallise and granulated honey
soon starts to frost. Bottling too far ahead always leads to
disappointment.

It seems hardly worth mentioning that the bottles need
to be clean and that they seldom arrive clean from the
makers. Bottle washing to remove dirt and film need not
be too tedious a process: use two deep sinks or bowls, one
with detergent and a bottle brush, the other with plain
hot water for rinsing. After washing and rinsing the
bottles, place them upside down on a wire tray to drain;
after a few minutes, tilt the tray to allow the drop which
accumulates in the centre of the bottle to run down its
side. Avoid wiping the jars with fluff-shedding cloths. If
the jars are examined, a raised ring will be seen just under
the glass screw threads. That is the mark which indicates
the right volume of honey to give the correct weight.

Raise the bottling tank onto an empty shallow super
covered with a board and open the tap gate slowly to run
in the honey, closing it just before the honey reaches the
shoulder of the jar. Then fill it with a trickle to the full
mark. Allow the drop to fall from the tap and slide a
clean jar forward to catch the next drop while you screw

on the cap. The necessary skills and knacks to obtain clean results come quite soon.

Stored honey granulates, and when the time comes for bottling it must be warmed sufficiently either to liquify it or to soften the crystallised mass enough for it to run. Resort once again to the warming box; 12 to 24 hours usually suffices to soften crystals to a semi-fluid which can be stirred and poured into the bottling tank and a 48-hour stay in the box usually liquifies the honey completely. An insulated box, controlled by a thermostat with a range from about 27°C to 50°C can be built by the enthusiast, but avoid using extended polystyrene as the insulator; glass wool is safer.

A better way of bottling granulated honey is to 'seed' the liquid with about 10% of fine crystallised honey kept for the purpose from a previous batch, and stirred into the warm liquid with a clean batten. After filling, put the bottles in a cool place where the honey will set in about a fortnight.

Honey for sale

When honey is sold in England from premises other than the producer's own home, a number of laws control its preparation and presentation. The bottles have to be labelled. If the word 'honey' appears on the label, no admixture of any substance other than pure honey is allowed. The label must carry the address of the producer, the packer or a selling agent, the weight must be stated, the honey may be sold only in approved packs of 1, 2, 4, 8, 12, or 16 oz or multiples of 1 lb, and the weights and scales used for checking the bottles have to bear an approved stamp. These amounts have not at the time of writing been affected by metrication in Britain, but may be in the near future. Other countries have similar regulations to which it is advisable to conform. The label must bear the name of the country of origin of imported honey. If it is sold in its home country it is always a good selling point to stress the fact.

Heather honey

European ling and Australasian manuka honeys have to be pressed in a heather honey press. The foundation can be saved by scraping the honey away from the septum into a scrim cloth with a spoon or loop of wire. The ends of the cloth are folded over and the cake placed in the press. Screw the top down slowly so that the honey can run through the cloth without bursting it. When the screw is down as far as it will go, release it and remove the wax cake, keep the cloth the same way up when refilling it for the next pressing. If the honey is intended for storage, run it into the can, but if it is to be bottled, tip it straight into the bottling tank, and stir it before running it off into the jars. As it has been pressed through a cloth it should not require further straining. Since the honey contains quite a lot of air the bottles will need filling about halfway up the neck of the bottle to give full weight.

Cut-comb honey

Instead of extracting all the honey or pressing heather honey, supers can be fitted with extra thin sheets of foundation without wires and consumed or sold as cut-comb honey. It is no more difficult to obtain than honey for extraction, but the frames have to be fitted out with new foundation for the next season and, of course, the bees consume honey to draw out the comb. To this extent the crop may be somewhat reduced. But comb honey, for the connoisseur, is so much preferable to extracted honey that the small beekeeper may well wish to concentrate on this method of producing honey until he or she feels that the expense of an extractor is justified. Honey was produced and eaten for many years before the first extractor was invented.

Collect supers in the usual way and store them in large plastic bags, sealed with Sellotape until required. Small plastic containers holding about half a pound of comb can be obtained from appliance dealers, as well as special cutters to cut and pick up a piece of comb to fit into the boxes. No expensive tools are required for small quantities: only a clean metal or plastic tray, big enough to hold a comb, a pail or honey tin, a palette knife or spatula, a sharp knife and two strips of wood. One of the strips should be as wide as the long dimension of the container and the other as wide as the short dimension, and both are made about 45 cm long. These can be used as knife guides for cutting the comb into slabs of the right size to fit the container. The pieces are picked up with the palette knife, slid off into the plastic boxes and pressed down lightly with a block of wood. The left-over pieces are then scraped off into the tin, together with any honey that has drained off, leaving the tray ready for the next comb. If these combs are cut too far in advance of needs they may granulate, but the uncut, stored combs seldom do. The broken pieces are treated as cappings after they have been mushed up or warmed overnight and strained for run honey.

Sections

Inducing bees to work sections is a great art in northern climates, one which demands the utmost skill of the beekeeper and some willingness on the part of the bees and the weather. The very thin foundations, put into sections, provide the bees with little spare wax for drawing out the cells, so they are obliged to produce wax in situ, instead of using the surplus from the thicker sheets normally used. But the diminutive size of the boxes prevents the wax-producing bees from clustering in sufficient numbers inside the partitions to raise the necessary heat, unless the weather is very warm indeed. The bees quickly show their dislike for section work if they are offered the alternative of the larger super frames by ignoring the sections altogether. The conditions for coaxing bees into supers of sections are precisely those which force swarming by congestion. The bees should be crowded into a small brood box, and the section super, placed above for drawing out and filling, does not provide sufficient space for all the adult bees.

Some bees seem to work sections more readily than others, and I rely on my bees to tell me at the beginning of each season which colonies will and which will not do so. After the spring examination I super four likely colonies with section supers and give them each about 2 litres of syrup to allow them to draw out wax. I know that two to four weeks later I shall have to increase the super capacity. By looking down into the sections I can see which hive seems happiest to go on working them, and I remove all the section supers from the unwilling bees, piling them onto the willing ones and putting ordinary supers on the others. I refer here to section supers, by which I mean ordinary supers filled with special hanging frames to hold three or four sections each, and I have deliberately refrained from using or advising the old-fashioned 'section crates'. These are special boxes designed to carry sections only, but I do not consider them to be a sound investment since they can be used for one purpose only. Section supers, on the other hand, can always be filled with standard honey frames if the season turns out to be unfavourable. It has to be admitted that perfect, well-filled and sealed sections can only be obtained during prolonged honey flows.

Sections, to look fine and clean, should be removed from the hive as soon as they are filled and sealed. The hive floor and interior must be kept scrupulously clean throughout the season. If the capped sections are left on too long, the cappings soon acquire a brownish tint, or, in beekeeper's jargon, they become 'travel stained', partly from being varnished with propolis, partly from being trodden over by bees running across the surface. A well filled, clean section weighs approximately a pound and is a delight to look at—and to eat.

Honey shows

Beekeepers' associations try to raise the standard of honey presentation by arranging annual shows for honey and hive products. There is no reason why your bees should not produce honey just as fine as your neighbour's, and if you take trouble with bottling you may win a prize even at your first effort. Don't be discouraged if yours is a third prize instead of a first. I have judged at many shows and in common with most other judges seldom find it easy to place first, second and third; the deciding factor is quite often some trivial oversight on the part of the exhibitor. But even if you never win a prize, which is unlikely, you will at least learn how to present honey in a pleasing way.

First watch the grading into light, medium and dark. When extracting, examine the combs through a strong light before you uncap. The distinction between the colours soon becomes obvious. Extract the light combs first, then the medium and last the dark. The keen exhibitor will even go to the lengths of uncapping only the light-coloured patches of cells on a comb and leaving the darker colours uncapped, for later extraction. When the honey has been bottled into spotlessly clean, selected bottles, screw down the caps and set the bottles in the warming box at a temperature of about 45°C for 24 hours, so as to clear the minute bubbles which cloud the honey. Do not exceed 50°C for fear of darkening it. Appliance dealers supply special grading glasses for honey bottled in standard jars. They consist of two yellow-brown glasses, one light and the other much darker. To use them, place a bottle against a plain white, well-lit background and hold a grading glass beside it. If the honey appears lighter than the light glass it can confidently be graded 'light', but if it is darker than the dark glass it goes into the 'dark' category. Honey colours between these two make up the 'medium' class. For crystallised honey choose light honey with a fine grain. The slightest speck of dust or trapped bubbles, or lack of clarity in the honey will downgrade the exhibit.

Heather honey is in a class by itself and should be as near pure heather as possible. Because of the bubbles, it weighs lighter than other honeys and tends to float in any mixture. A good sample can usually be obtained from a storage tin by stirring the upper half only and ladling from the top into a few jars. From the jars bottled, select the best after careful examination for flaws or specks. At the show take with you a box of new selected screw caps, polish the jars and replace all the caps before setting up your exhibit.

Sections need to be full, except for the last ring of cells nearest the wood, free from travel stains and, when held to the light, equally tinted throughout with no dark pollen cells; the wood needs to be cleaned of propolis and boxed as required by the schedule. A class for a comb of honey for extracting needs only a careful selection, at extracting time, of your best, clean, well-filled, evenly flat comb presented in a glass-sided case. Scrape the frame free from propolis and brace comb, and make sure that the comb cannot swing in its show case and bruise the cappings. Some exhibitors tighten a thumb screw through the case into the side-bar to hold it firm. If any cells are damaged when cleaning, the comb can be replaced in the super for the bees to clean up overnight.

Whether you exhibit or not, you will gain a lot from visiting a honey show and learning how to present honey, and from talking with other beekeepers. Remember that the judge is not really giving marks and prizes to the honey, only to the beekeeper's work in presenting it. A small beekeeper with only one or two hives can devote just as much or more care to his exhibits than the busy commercial beekeeper who is prone to select a few jars from his shelves with no special preparations at all.

Wax

Wax is produced at the expense of honey; it costs quite a lot to buy and every scrap should be saved and rendered into clean wax cakes. These are exchanged with appliance dealers at fixed prices, for foundation or other goods, or it can be cast into foundation in do-it-yourself moulds. Render cappings and old comb as soon as possible, otherwise the wax moth will certainly destroy them within a very short time. Wax is rendered either in a solar wax extractor or by heating in water.

The solar wax extractor

This costs some money or labour in the first instance but costs nothing to run and is surprisingly efficient. Properly made it can melt wax in any month of the year, provided that the sun shines. Odd pieces of wax, cappings and old combs stood vertically are placed under the double glazed top and a cake of wax is retrieved at night. The dross left in the tray can be used in the compost heap and is a good garden fertiliser. The cakes are put through a second time, with a piece of old flannel or cotton lining the tray, and raised over the straining bar. This second cleaning produces high-quality, clean wax, but avoid loading the tray with more wax than the mould can contain.

Wax boiler

Efficient wax extractors work on the principle of boiling or steam-heating the old comb and pressing out the wax through sack in a heavy screw press under hot water. Only a large enterprise could justify the expense of such a piece

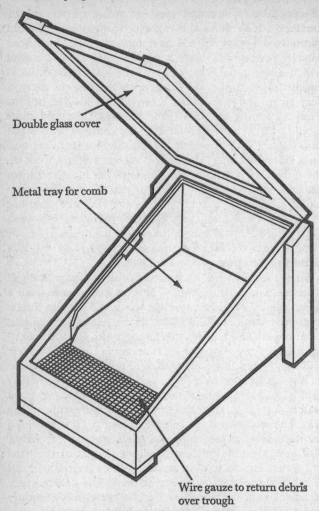

Double glass cover

Metal tray for comb

Wire gauze to return debris over trough

FIG. 30 Solar wax extractor

of apparatus. The small wax extractors offered to the amateur seldom have the capacity to deal with the old comb from a single hive. I have for some years now been working exclusively with a simple and efficient device which costs little and leaves very little wax in the dross. It consists of a 5-gallon oil drum with the top cut away, a very large, fine-hole conical sieve with a handle, bought from a catering supplies firm, a metal ladle and some moulds—loaf tins are ideal. The old comb or wax is put into the can with rain water and brought near to the boil on a kerosene boiler; the mass is then stirred to break up the comb and loosen the wax. The strainer is pushed down into the mixture and the floating wax in the strainer is ladled off into the moulds to set. After several repetitions to remove as much wax as possible, the dross is left to cool. Only a thin layer of wax remains on top; this skin can be removed and set aside for the next boiling, and the bulk of the dross or 'slum gum' composted.

Since fuel oils have become so expensive, I have changed from a kerosene stove to a U-shaped grate made from a dozen bricks. The fuel is garden waste, old broken frames and workshop off-cuts. I find that I can remove the wax from old frames, boil up the combs and burn the broken frames all at the same time. At the end of the day I am left with cleaned frames and cakes of roughly rendered wax.

Wax melts at or near 62°C and begins to darken fairly rapidly if held too long at the boiling point of water. Hard water spoils it to some extent by saponifying some of the wax, and for this reason soft water or rain water is used for the boiling. A tablespoonful of vinegar to every gallon will neutralise the calcium for those who must use hard tap water. On no account should wax be heated without water, and pans must not be more than three quarters full. If wax is required for show purposes, use only cappings and pour the strained wax into moulds with some clean boiling water at the bottom. When cold, scrape away all impurities before the final casting. The art of making a

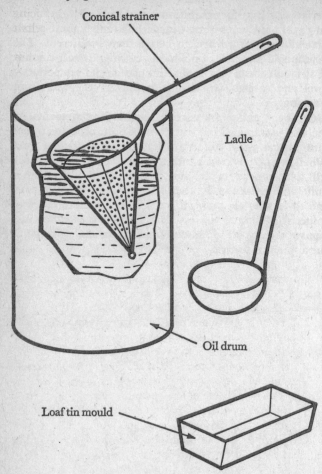

Conical strainer

Ladle

Oil drum

Loaf tin mould

FIG. 31 Equipment for extracting wax

perfect wax cake for exhibition is best learned by inducing an expert to demonstrate his method, and I must admit that my own efforts are no better than indifferent. The final mould has to be of polished glass; the correct amount of strained, clean wax ladled in; the mould set perfectly level and cooled very slowly to prevent the wax from cracking.

Ordinary grades of wax require no such refinements. As the wax is skimmed off, a certain amount of water comes into the ladle and collects in the mould under the wax. When the cake is cold knock out the wax, and while it is still wet scrape away the loose dross under the cake with a knife and discard it. Scrape the thin layer of mixed wax and dross over the wax bin, so as to recuperate the wax content at the next boiling. The remaining wax cake can be used at once for foundation making, or for exchange at the appliance dealers.

9

Food and Feeders

Feeding bees

Just consider for a moment how bees cope with their own feeding problems. In the early part of the season they gather what they can to supplement their stores and consume both pollen and nectar nearly as fast as they bring it in. When the honey flow starts they set to, with will and purpose, to lay by huge reserves to last them out the rest of the year. While we may think that they gather their stores over a period of six to nine months, in fact they gather the greater part of their year's supply in a very few days or weeks. We must conclude, therefore, that bees are well able to deal with large amounts of food at a time, and that regulating a small trickle of food into the hive, over a long period with repeated visits, contributes little or nothing to the well being of the colony. When bees need feeding, they need reserves sufficient to carry them over to the next flow, be that a week or six months.

Why feed at all?

It may well be asked why bees should need feeding at all since they supply us with surplus honey; surely they should also have enough for themselves. This is perfectly true, and I for one try to leave bees an ample supply of their own natural food for their winter supply, but the matter involves complications which go beyond good intentions.

It frequently happens, particularly with stocks that have swarmed or have been artificially swarmed, that the queen goes on laying right up to the middle or end of a flow. The eggs continue to develop for a further three weeks, before the bees can emerge and vacate their cells in the brood nest. While the flow is in progress, the bees store in the only cells available and these are placed above the queen excluder, particularly with small brood box hives. By the time the brood cells have been vacated, the flow has ceased and they remain empty. The beekeeper then removes the 'surplus' and leaves the colony with practically no stores to face the winter. Under these conditions, the bees need feeding either by taking them to a late flow area, such as the heather moors, or by feeding them sugar to replace the stores that have been removed.

In general, whenever we tamper with the normal cycle we disturb the bees' foraging ability, and they need feeding to put them back into a satisfactory state for their further development. Some processes, such as rearing queens, only take place naturally during a flow, and feeding has to precede and accompany any artificial queen rearing we wish to undertake. The following list will give some idea of the circumstances when feeding becomes necessary.

Installation of a swarm or package
The bees have to make wax, for which they need to consume honey before they can begin to make a nest and store for the needs of the brood and to last out the winter.

Making a nucleus
A nucleus has few bees, and although it has been given a reserve comb it will need food to draw out new combs and conserve the energies of its foragers for 'brooding'.

Putting on a super of foundation or sections
This should be done well before the flow; the bees will need the food to draw out the cells. Of course, never feed

during a flow or the bees will store the sugar syrup in the supers, after the flow, and that will spoil the honey. There is no need to feed when supering with drawn comb, unless the stores in the brood box are exhausted.

For queen rearing

Although only simple queen rearing has been covered in this book, it is well to bear in mind that good queen cells can only be reared when a net increase in the stores is taking place. Some feeding should accompany every stage of queen rearing.

Whenever stores are getting low

A check needs to be made from time to time, both in warm winters and in cold or very dry summers, to avoid disastrous checks to the normal cycle. Bees consume more in mild winters than in cold ones because they fly more and expend more energy. In cold or dry summers, increasing numbers of mouths to feed deplete stores with alarming speed, unless there is some income. Halfway through the winter and after the period of maximum brood are always critical times.

What to feed

At first sight it would appear that the answer to this is quite simple: honey. Alas, feeding honey to bees is fraught with dangers and should never be contemplated unless the honey is in combs, preferably from the same hive or from the same apiary at least. Honey may carry the spores of the worst of the bee diseases and spread them. Liquid honey supplied in a feeder tends to cause serious unrest in the apiary. Honey in the comb is quite safe because it becomes part of the brood nest without spreading its aroma to attract potential robbers. Above all never feed imported honey to your own bees. In many exporting countries, foul brood disease is held in check by antibiotic treatment which prevents development of the pathogens but does not kill those which may be present in honey already stored. A number of outbreaks of disease can be traced back to the kindly beekeeper who buys a container of honey at the supermarket to feed to his bees 'to tide them over'.

The only food, apart from pollen substitute, which should ever be given to bees is white sugar syrup. Many of us, concerned at the growing over-purification of our human foodstuffs, try to restore the balance in our diet by retaining roughage, natural salts and mineral impurities, preferring wholemeal bread to white, and brown demerara to white purified sugar. I do so myself, and I prefer natural honey to either—but bees do not have human stomachs, nor can they cope with the kind of impurities which arise in the boiling of sugar cane juices and beet pulp. These products of boiling are not entirely natural and cause dysentry to wintering bees.

There need be no fear of reducing the vitality of the bees by a white sugar diet. Their period of growth and development occurred during the five days of larval feeding on milk elaborated by the nurse bees. Provided that there was a fair amount of honey to start with, the small admixture of sugar can have no deleterious effect. The same is not quite true of pollen substitutes, where second and third generation bees raised on artificial diets do seem to suffer somewhat, probably through the poor quality of milk fed to the queen during her heavy laying spells. Adult bees use honey or syrup only as fuel to supply their energy requirements.

Some old beekeepers still like to feed 'candy'. This form of sugar has no advantage at all over syrup at any time of the year, and bees take it so slowly that they are unable to develop with its help. If they need feeding at all, feed them syrup at the right time to store, concentrate and seal, or feed to stimulate rapid development in spring. Emergency winter feeding can be undertaken with syrup in a contact feeder placed on the frames directly above the cluster.

Economics of feeding

At one time, it was thought that if we removed 20 lb of honey from the brood box and replaced it with 20 lb of sugar made into syrup, we should be better off by twenty times the difference in price between sugar and honey. When we deal with animals, the food conversion ratio is never as perfect as that, and bees are no exception. When we feed in the autumn, a fair amount of energy has to be expended to bring about conversion of sugar to low grade honey and further, misled by the untimely 'flow' from the feeder, the bees will induce the queen to lay, and a size-able proportion of the incoming food will be used to pro-vide heat for the brood nest. If we then add in the cost of fuel to make the syrup and our time and labour, perhaps even travel costs for our out-apiaries, we shall soon find that sugar feeding, at present price levels, does not com-pare very favourably with leaving the bees their own stores. I hold strongly to the view that stores in the brood chamber 'belong' to the bees and that I should only feed to supplement their natural reserves in a brood chamber large enough to hold the winter's full supply. In the British context, the 'winter' may well extend into June!

How to make syrup

In the autumn we want our bees to take down and store the maximum of sugar, converted into glucose and fructose. For this reason we feed them heavy syrup, and in every case where the aim is to replenish stores, heavy syrup will be needed. The usual recipe for this reads: take 2 kg of sugar to 1 litre of water, bring to the boil and stir until all the sugar is dissolved. This involves measuring and boiling which is quite unnecessary in practice. If we put into a container the amount of sugar we wish to feed, level off the top and mark the level with a soft pencil, we then add boiling or very hot water, stirring as we pour, until the syrup reaches the original mark. It will cool, ready for use all the sooner, because the solution absorbs heat. Any kind of container can be used, but it must be borne in mind that if sugar syrup remains for long in aluminium or zinc vessels a reaction takes place which produces compounds slightly poisonous to bees and which eat through the metal. Tin, wood and plastic are quite safe.

For spring and nucleus feeding, we do not usually need to provide for long-term stores, but to supply food for current consumption. The bees will be able to consume the thinner syrup directly, without being obliged to go out of the hive to fetch water to dilute it. For spring stimulation, we feed in the proportion of 1 kg of sugar to 1 litre of water; this is obtained by marking the level of sugar in the vessel and then gauging by eye or ruler a quarter of the height of sugar. We then make a second mark, one quarter higher than the previous one, and fill with water to that level.

Cold water syrup

Heating is not always possible in out-apiaries, but syrup can still be fed in certain types of feeder. The simplest of these is a lever-lid tin, such as a cleaned empty paint or emulsion can, or a large Nescafe or honey tin. About a dozen holes 1 to 2 mm in diameter are knocked into the lid with the point of a round 35-mm nail. For heavy syrup, fill the tin to within 2 or 3 cm of the top, then pour in cold water, stirring with a stick to dislodge the trapped air, and fill to within 1 cm of the top. Press the lid down all round and invert the tin over the feed holes or directly over the frames. For thin syrup, fill the tin to two thirds of its height, but bring the water up to 1 cm as before. When the tin is inverted, the heavy saturated syrup falls to the bottom where the bees can reach it and gets replaced as the water dissolves more sugar. The small amount of spillage on inversion will serve to attract the bees to the holes. Not more than a dozen holes should be made, to prevent the syrup being taken down more quickly than the sugar can be dissolved. This method of feeding has the advantage of cheapness in the cost of the feeder and saving in fuel, but tends to feed more slowly than with some of the better known feeders. If old paint or emulsion tins are used, check first on the label that the contents were not dosed with insecticide or fly-killer chemicals, and discard any that have been so labelled.

Feeders

Many types of feeder are currently on the market, several of which are good and some of which are useless. Feeders holding less than 5 litres can be used for nuclei and small swarms. When placed over large hives they needlessly multiply the number of visits required. A reasonable feeder should be able to hold 10 litres of syrup: enough to supplement the winter feed in one operation. Aluminium has proved to be unsuitable as a material because of sugar corrosion, and tin, although harmless, has a tendency to rust under normal conditions of use, thus limiting its effective life; this can be overlooked in the case of reused tins because they cost little or nothing to start with and are easily replaced. The best materials for any type of feeder are wood and plastic. Feeders can be grouped according to the way they operate.

Contact feeders

The lever-lid tin, previously described, with holes pierced in the lid, is cheap, efficient and can be inverted directly over the combs for feeding in cold weather. Sacks or old cloths cover the combs temporarily while the feeder is in position, or it can be placed over a cover board with central feed hole. Plastic buckets with wire cloth embedded in the lid provide an excellent and permanent version of the contact feeder and are now listed by most dealers. These feeders do suffer from one great inconvenience: they can only be used inside an empty hive box, either brood chamber or super, in order to make the hive bee-proof. Empty hive boxes seldom occur in a rational apiary. Each box should have its complement of combs in use, or ready for use, and removing the combs to leave space for the feeder exposes them to serious risk of damage. I would prefer to make a temporary box of packing case wood than to tip valuable combs on to the workshop floor —combs are the most valuable asset the beekeeper has, after the hive and the bees.

Trough feeders

The simplest way to feed syrup would be to provide a trough or bowl and just let the bees come and take what they need, but a large expanse of liquid food, with its attractive smell, can easily lure many bees to death by drowning in the syrup. Precautions have to be taken to ensure that not more than 12 mm width of surface is available to them; then, if they do fall into the liquid in their eagerness, they will be able to reach the side very quickly and to crawl out before they drown. All the many different patterns of trough feeders are designed to that end. The larger round feeders of biscuit tin size have a central hollow column reaching to 8 mm below a box surrounding the column with a 12-mm space between the box and the sides of the column. A lid covers the tin to prevent the bees from entering the main body of the syrup. Provision has to be made for the liquid to flow under the box, and it is usual to surround the column, inside and out, with perforated metal to give the bees a foothold. Do not omit any of the parts when feeding. It is a most distressing

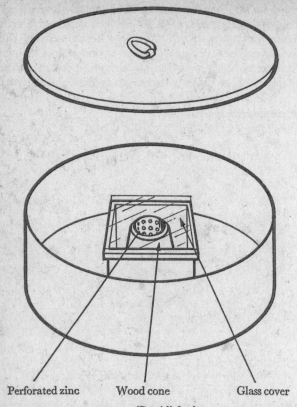

Perforated zinc   Wood cone   Glass cover

FIG. 32   'Rapid' feeder

sight to find the entire surface of the syrup covered with the bloated corpses of drowned bees. Many appliance dealers now have a plastic version of these so-called 'rapid feeders' with a capacity of 7 litres, but these, too, have to be used in a hive box, and the central plastic column has to be sand-papered rough or surrounded with perforated zinc because its glossy surface does not give wet bees a foothold.

By far the best feeders to have are the 'overall' type, known in the trade as Miller feeders, which give central access to the syrup. Another even better version, called the Ashforth feeder, gives access at one end. The overall feeder consists of a trough of the same perimeter dimensions as the hive and about 7 cm in depth. The syrup is

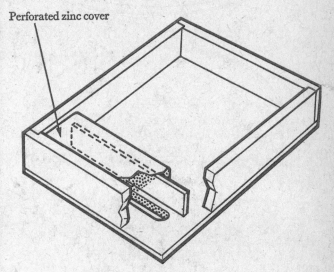

Perforated zinc cover

FIG. 33   Ashforth overall feeder

restricted to one or two troughs with a channel running full width. The access wall rises some 8 to 10 mm lower than the outside walls and a bee-proof cover allows a long narrow approach to the food. A top cover is unnecessary if the roof is correctly bee-proof. When placing it on the hive, be sure to see that access is available at the lowest point of the slight forward slope. In the case of the Miller feeder this will mean that the channel runs from front to back; the Ashforth channel has to be placed parallel with the hive front. Some of the older makes of overall feeders have metal trays for the syrup, but these rust badly and deform sufficiently for bees to get into the main trough. The 'do-it-yourself' enthusiasts can make these overall feeders easily enough if they remember to glue the parts together with plenty of thickish cascamite to make them waterproof and then paint the inside with several coats of a good bituminous paint.

Plastic bag feeders

This novel once-only type of feeder can be used when the cover board allows the bees access near the corner or perimeter instead of through a central hole only. A square 'ring' to fit the hive and about 5 cm in depth has to be made to raise the roof high enough. This can be made by nailing together four pieces of 16 or 18 mm × 46 mm wood cut to the perimeter dimensions of the hive. I use a large, thin polythene bag about 35 cm × 50 cm and place it in a bucket, also of polythene, open out the bag, and pour in a bucketful of syrup. I then take up the open end of the bag and slightly raise it in the bucket, twist the unused part, expelling any surplus air, and tie the roped material into a tight knot. If there is a central feed hole I cover it with a piece of slate, glass or plywood and lay the bag on the cover, put on the 'ring' and allow the bag to settle for a moment or two. With a 5-cm nail or pointed match stick I make about a dozen holes in the flat top of the bag, always on the upper surface, and press the bag to bring some syrup through the holes. The bees work these bags quite quickly; the thin polythene collapsing as the liquid level falls. The bag feeders cannot be used with standard central feed hole covers. Ordinary inner covers can be drilled with an 18-mm hole at each corner, but these have to be covered or plugged if the board is used to clear supers with a Porter escape. The New Standard cover board with its perimeter ventilation is ideal for the purpose.

Frame feeders

With large brood body hives, such as the Jumbo or the Modified Dadant, excellent feeders can be made, but

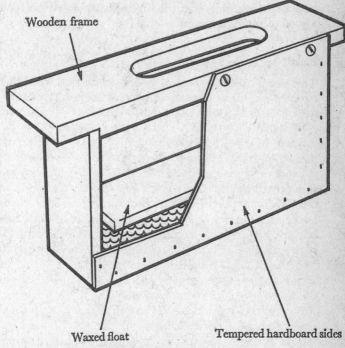

Wooden frame

Waxed float

Tempered hardboard sides

FIG. 34 Frame feeder

seldom obtained from the appliance dealers, with narrow deep troughs, of the same external dimensions as two frames. The surface of the liquid is covered by a waxed or varnished wooden float. I use these with great success in Jumbo hives as dummy frames. If the brood should require the full ten combs for its development, the feeder can be removed and two frames substituted. The handy beekeeper can make these quite easily with wood and oil-tempered hardboard. A most useful feeder devised and used at Sparsholt by Captain Tredwell rejects the float and allows access by the bees to two narrow perforated zinc channels in which they cannot drown, whatever the height of the syrup. These feeders will last for many years and can be filled with syrup by sliding the cover board over to reveal the filler hole. It is wise to be systematic about the placing of the feeder always on the same side in all hives, so that you always know which way to slide the cover board to reveal the feeder. In general, hives smaller than the Jumbo have too little comb space to allow the use of these frame feeders, but a one comb width version is invaluable for feeding nuclei and small or weak stocks. The Jumbo or MD feeder holds nearly 10 litres and, if desired, a Langstroth frame size has over 7 litres capacity. Smaller, single-frame feeders are ideal for nucleus hives.

I am often asked how many feeders are required for an apiary. A feeder is an essential part of a hive, and one per hive with additional smaller feeders at the ratio of one per nucleus hive should be considered the norm.

Emergency feeding

If, through some neglect on the part of the beekeeper, a colony is reduced to starvation level, immediate short-term measures need to be taken to revive or bring the bees into condition to take down the syrup from the feeder. Syrup can be sprinkled into the cells of empty combs near the cluster. Remove the cardboard wad from a honey jar, knock about ten holes, 2 to 3 mm in diameter, through the tin lid with the point of a nail, and shake the syrup through the lid on to the horizontal comb face after shaking off the bees. Replace the comb or combs on the hive and give a normal feeder. It is a good idea to trickle some syrup through the bees' access so as to attract them rapidly to where the food is. When working in out-apiaries it is always advisable to carry a can of water and a towel to remove the stickiness from your hands before handling the steering wheel of your car.

Bees' food requirements

Some strains of bees need more feeding than others. Generally speaking, our native or mongrel dark bees are less prolific than the more highly bred Italian or Caucasian races, and their breeding continues later into the flow. When bees are less prolific the space available in the brood chamber for storing is greater. As there are fewer bees they have less mouths to feed and less bees to gather surplus. For the amateur, these bees tend to give less trouble, less work and less honey. On the other hand, the prolific bees use all their brood space for rearing young bees, store more in the supers and when the flow comes have a large force of foragers ready to fill combs for the beekeeper. When feeding is resorted to, they tend to resume breeding once again and less of the food is stored. The management system has to take these factors into consideration. With greater crops as a reward, the beekeeper has to work harder and lay out more money in sugar to earn the extra 'take'. It is up to each one to decide for her or himself which kind of bee to keep in relation to the work involved. I personally like the more prolific strain and adjust my management accordingly, but I can well understand those who prefer the thrifty, less prodigal bees and content themselves with more modest crops.

Open-air feeding

In some countries, pails of syrup sprinkled with granulated cork floats or straw are placed about the apiary to allow the bees to come and gather their own food. In isolated apiaries, with all the colonies of even strength, this method offers advantages of speed and simplicity. But these conditions are seldom met with near my own apiaries; within a relatively short distance I have neighbours who also keep bees. Whilst I bear them no feelings of ill-will, I still do not think it is up to me to bear the cost of feeding their bees as well as my own, and the open feeders would be sure to attract every bee within several kilometres. I do try to keep my bees' development as even as possible, but at the end of the winter there are some stronger than others, and individual management helps to straighten out the apiary. But if I should use open feeders, the stronger colonies with a larger foraging force available would soon clean up the food while the weaker stocks would have a proportionately smaller share; the exact opposite of what is required.

Feeding and robbing

Quite a lot of nonsense has been and is being written about the dangers of robbing when feeding bees. Feeding bees inside their own hives with sugar syrup causes no disturbance in the spring, provided that the feeders are sound and the hives bee-proof. Obviously, a leaky feeder, laying a trail of trickling syrup from the hive to the ground, is going to attract unwelcome visitors. Test the feeders before use, by filling them with water and watching for leaks. These can be stopped with quick drying plastic fillers such as 'Plastic Padding', but this should only be necessary with old or neglected equipment. However, if outdoor feeding is resorted to for any reason, do not remove the feeders until the bees have emptied and abandoned them. Exactly the same is true if pieces of honey-laden comb or tins are being cleaned out by bees. Once the bees have found a source of food, robbing will start if it is suddenly removed, but no trouble will be experienced once every trace of honey has been cleaned up. The reason is not far to find. Bees are well accustomed to flows coming to a natural end; as the forage diminishes, more and more bees will turn their attentions to other sources or duties. The honey or syrup left in the apiary acts as a minor flow. If this 'forage' is suddenly removed before it has become exhausted, all the bees, engaged in cleaning up, will search around to locate it. If they find the entrance of another hive from which issues the looked-for scent, then the frustrated gatherers will certainly try to overcome the resistance of the guards and rob mercilessly.

10

Health in the Apiary

Hygiene

Bees produce honey and honey is food for human consumption. The apiarist has a duty to the consumer to see that all the hives, combs, feeders, tins and other equipment are kept scrupulously clean. Any honey or syrup fed to bees can be, and often is, moved by them out of the brood chamber and up into supers. For our own sakes let us help the bees keep their homes and food stores entirely free from contamination. The bees do their best to keep their nest clean according to their own instinct, and we should ease their task rather than make it more difficult. But we are more concerned in this chapter with the health of the bees.

Poison sprays

The modern farmer or gardener protects his crops from insect damage by the use of insecticides in the form of dusts, granules and liquid sprays and many control weeds with herbicides. Most of these affect bees as much as they do other insects. The substances used are strictly controlled in most countries and toxicity to useful pollinating insects assessed. Growers are advised about the precautions to take to avoid damage to humans and beneficial wild life. Unfortunately these instructions are often forgotten or ignored and many bees and even whole colonies are destroyed by thoughtless spraying. Damage to bees occurs in at least four ways:

1 When bees fly through a toxic spray cloud on their way to their foraging area.
2 When open flowers are sprayed with vaporised liquids while bees are working them.
3 When toxic dusts are blown on to open flowers. The dusts are collected by foragers with the pollen and stored in the combs, where they may remain toxic for months.
4 When beekeeping equipment, foundation, etc. is kept anywhere in a building where 'Vapona' or similar resin bound 'dichlorvos' flykilling paper is used. The fumes can cause enough contamination to kill bees up to several months later.

In addition, it is worth noting that dusts are more harmful than liquid or sprays and that granules do least harm to bees. Unfortunately, granular formulations cannot be used with all crops since they depend for their effect on lodging in the leaves, axils or stalks and releasing poisonous fumes over a long period. Plants which fail to trap the granules prevent the more frequent use of this least harmful form of insecticide.

Some herbicides also kill insects, but their effect on bees can only be felt when a honey or pollen-yielding crop is chemically killed during the flowering season. If a field is sprayed with herbicide to kill off the weeds after a cereal crop, many bees will die if composite (daisy family), cruciferous (radish family) or other flowers are being worked. Fortunately, such late season flows seldom involve more than a small proportion of the bees at any one time, and spray damage passes largely unnoticed by the beekeeper.

Beekeepers tend to be too fatalistic or apathetic about the disasters which follow spray poisoning, and many have suffered losses that might well have been avoided had they been more active in taking protective measures. In the first place, a beekeeper has to make himself aware of the major flowering crops which are liable to be sprayed within a radius of 3 km from his apiaries and to realise the potental danger. If such a crop is grown nearby, the beekeeper should at once make himself known to the grower and seek his co-operation in minimising harm to the bees. It is surprising how co-operatve farmers can be when they know that they are putting another man's live-stock at risk. They are always very helpful in words, but co-operation may be less apparent in subsequent deeds. Most farmers spray in a last-minute panic instead of calmly at the right time.

When bees are poisoned

Since the major damage to bees occurs when flowering crops are sprayed in fine weather during the daytime, common sense dictates that operations conducted in the early morning, late evening or night, or before or after flowering instead of at full bloom, or during dull cold weather, will reduce damage to bearable proportions, pro-vided always that the insecticides are non-persistent. It is significant that most of the spraying which harms bees is also uneconomic from the point of view of the grower. If he applies insecticide when the flowers are open, the pests have already done extensive damage to the crop, and if he kills his weeds when they are flowering, it is far too late; the weeds have already reduced his crop. May we hope that one day, either voluntary or enforced enlightenment will drive home to farmers that protection of their crops is quite compatible with the preservation of pollinating insects. Beekeepers will no doubt have to play their part in informing farmers where their joint interest lies.

Emergency protection of bees

Many people pin their hopes of protecting bees on warn-ing schemes which would allow the beekeeper to close up his or her hives during spraying operations, but if hives are closed up during fine days the bees are in jeopardy, so the warning merely tells the owner that if his bees fly they will be poisoned or if he closes them up they will suffocate. There is no certain way of closing hives in warm weather without endangering the whole colony. The dangers can be minimised by careful consideration of the factors involved.

If bees can see light at the hive entrance, they will try to get out and block the normal ventilation which would keep them cool. The enforced crowding of the bees over the combs causes a catastrophic rise in temperature which kills brood and can even break down the combs by melting the wax. A palliative therefore would be to place an empty super or brood box between the floor and the brood and to block the entrance with wood or foam plastic, put a contact feeder full of water directly over the frames of the top super, remove the cover board, and tack a piece of hessian or fix a screen over the box surrounding the feeder. The roof should then be placed askew over the top to allow for air circulation. With these precautions the colony has a chance of surviving for a period of twenty-four hours.

A reasonably effective alternative is to put a feeder of water on the hive and literally bury it in a loose covering of hay or straw about 30 to 40 cm thick. Some bees will come out, but the obstruction will so disorientate them that they will not fly away from the hive for the first twenty-four hours. All these protective measures must be taken the night before the spraying is due and will not guarantee complete freedom from harm, although losses will be reduced and often eliminated.

Productive but dangerous crops

Most fruit grown for market benefits by bee pollination and provides some sustenance for bees, and all such crops are nowadays protected by insecticidal sprays. If at all possible, permanent apiaries should be sited well away from large orchards and fields of nectiferous cultivated crops. Growers should be and sometimes are willing to pay the beekeeper a fee to bring in bees for pollination services immediately the last pre-blossom spray has been applied and to have them removed before petal fall. Timing is all important and the beekeeper has to be prepared to move the hives in and out again at twenty-four hours' notice. The pollination fee has to take into account transport costs, possible losses through neighbouring farmers spraying after the hives have been opened, losses through disease caused by moving the bees and a fair charge for labour.

Field beans now receive applications of granular insecticides, which seldom harm bees, but by far the most dangerous crop for the bees is oil seed rape, at the same time the most profitable for the beekeeper if all goes well. Here again, it is best to bring bees to the crop rather than have an apiary sited near it, and to move the hives away a week or so later.

In Britain, the Ministry of Agriculture publishes annually a free 'List of Approved Products for Farmers and Growers', which enumerates the range of chemicals that can be used for spraying crops and outlines their toxicity to bees and persistence of the toxic effect. When talking to farmers about their spraying programmes as it affects your bees, it is invaluable to know what the dangers are by consulting this informative booklet.

Bee diseases

Bees are subject to many diseases. For practical purposes we can divide these into brood and adult diseases, all caused by some pathogen, virus or microbe which can spread rapidly in the hive and result in the death of the entire colony by destroying only the susceptible half. If the brood dies the colony eventually dies out; if the adults die, the brood starves to death. Some of the diseases strike at previously healthy and strong colonies and can only be controlled by destroying the pathogens or all infected colonies within flying range. Others arise in colonies that are put under stress, and pathogens which were previously held in check in a stable colony quickly weaken and destroy it under adverse conditions. Sickness caused by stress conditions is within the powers of the beekeeper to avoid to a very large extent. For some of the diseases, medication allows a measure of control, but as in humans there may be side effects that we know little about as yet.

Brood diseases

Sickness of the brood means that the larvae or pupae die in their cells before they reach maturity. Instead of emerging and vacating the cells at the same time as the other bees around them, the dead larvae are either removed several days or weeks before their time, or else their remains putrefy in the cells and prevent the queen from laying evenly in batches. This invariably results in a patchy brood pattern during the breeding season. The effect is described as a pepper box pattern because of the numerous empty gaps in the sealed brood combs. All brood diseases exhibit this feature to some extent, and the finding of such brood patterns must always arouse suspicion. Further inspection of diseased cells, sealed or unsealed, enables the experienced beekeeper to suspect which of the diseases has caused the trouble, but final diagnosis has to be confirmed by specialist microscopical analysis.

American foulbrood (AFB) is a disease of sealed brood. The pathogen of this extremely infectious bee sickness is called *Bacillus larvae* (White). It forms spores which are highly resistant to nearly all disinfectants, to great degrees of heat and to very long periods of desiccation, extending to several decades. Young larvae absorb the spores with their food, but show no symptoms until the cells are sealed over. During the prepupal rest period the bacillus invades the tissues, proliferates rapidly and kills its host. It then forms spores and can infect other bees. The capping over the infected brood sinks and acquires a dark wet appearance. The bees make some attempt to remove the cappings by biting out small pieces of wax. The resulting perforated cappings are quite easily distinguished and always need investigation. If at this stage a matchstick is inserted into the cell and stirred, the contents of the cell can be drawn out into a long glutinous thread or rope, about 2 to 4 cm in length. The ropiness test provides fairly conclusive evidence of American foulbrood, but can only be applied to brood at a specific stage of decay. The dead larvae or pupae collapse, fully extended on the cell base, adhering to it so strongly that house-cleaning bees are unable to remove the so-called 'scale'. On old infected combs, where the ropiness test cannot be applied, the comb can be held in such a position that the light falls direct into the cells. By looking down the comb, the raised desiccated heads of the scale pupae can be seen. The spores in such scales have been known to infect colonies for up to sixty years after the original colony died out.

Whenever a beekeeper suspects that he has AFB, he should, in his own interests, contact the nearest district bee disease inspector, who will carry out the necessary and obligatory prophylactic measures. The entrance should be restricted to prevent robbing, and hands, hive tool and smoker should be washed in running water before opening another hive. If the colony is already dead, block the entrance to avoid any further robbing. The whole hive, brood, bees, combs and honey are highly infectious. The treatment given varies from country to country, but British law requires that the bees be killed at night and the entire contents burned in a pit; the hive scraped and subjected to flaming with a blow lamp, all done by or under supervision of a bees' officer. In some countries the colony is treated with antibiotics which suppress the symptoms in the bees, but do not kill the spores present in the hive. As soon as such treatment is suspended the disease breaks out again and the honey remains infectious. In addition, some of the antibiotics used—streptomycin is one of them—can remain stable in honey for several months. Honey laced with streptomycin can be consumed by humans and affect their reactions to this antibiotic, which is also used in human medicine. Honey infected with any bee pathogen is quite safe and wholesome for human consumption, but the disposal of jars and containers presents some danger to nearby colonies.

Frequently the strongest colonies in an apiary are the first to succumb to the disease because they have large numbers of foragers which soon find any ailing stock within flying range, and rob it of its infected honey. Beekeepers' associations usually provide insurance for their members against losses incurred by enforced destruction of their bees due to foulbrood.

European foulbrood (*EFB*) is a disease of unsealed brood. It is one of the stress diseases, and bees that suffer from lack of food in the spring or are moved early to top fruit pollination are very prone to it. The causative microbe, *Streptococcus Pluton* (White), enters the young larvae with its food and proliferates in the mid-gut, which it quickly fills and blocks, preventing the absorption of further food. The larvae then decompose to a yellowish colour, losing the annular ring structure of their bodies, and acquire a shiny wet appearance; they lie in unusual positions in the cell, most frequently in a slightly spiral extended posture. If not cleared away by the house bees the larvae can dry into a kind of scale which is easily detached, unlike the scale of AFB. Strong, well-fed colonies can overcome the condition, or the advent of a honey flow may avert its development temporarily. Once the trouble has started the streptococci can remain alive and dangerous for up to three years in the comb, despite the fact that they do not form spores. A number of secondary infections attack the dying or dead larvae, but not until Bacillus Pluton has done its work. Infected colonies frequently survive an attack of EFB, but are so weakened in the spring that they seldom prove profitable. The disease can seldom be found in any colony after the main honey flow.

Treatment for an initial outbreak should entail killing and burning the bees, frames, combs, and honey and scorching the hives, just as for AFB. But in apiaries where the malady has become endemic, the bees can be cured by the application of Terramycin, TM25, in syrup for two or more successive years. In England and Wales, Terramycin may only be purchased and used legally under the supervision of a veterinary surgeon, but the bee disease officers will diagnose and treat affected colonies free of charge. After Terramycin treatment, the beekeeper would be well advised to make a complete comb change to rid the hive of the pathogen (see under 'Nosema' below), then melt down the old combs and scorch out the hive and frames. Streptococcus Pluton is destroyed by boiling wax, honey, etc. in water or by maintaining at 88°C for 8 minutes. Some districts and countries are completely free of EFB while others have a high incidence. Factors of climate and soil alone seem to account for this variation.

The two foulbroods are recognised in most countries to be the most infectious forms of bee disease, able to spread from hive to hive and apiary to apiary. The various governments have enacted laws to ensure regular inspection of apiaries and treatment of affected colonies. In England and Wales, for example, the Foulbrood Disease of Bees Order, 1967, provides for compulsory examination of all bee colonies at least once every three years and for compulsory treatment of infected colonies. Since the first act came into force in 1947 the incidence of AFB has been reduced to a tenth of its previous spread, and EFB has been contained in total numbers in spite of its recent spread to several new districts, mainly the large fruit growing areas, where bees have been brought in for pollination services.

If you have the slightest doubt about the health of the brood call in the local bees' officer. If the disease is present, it may be possible to contain the epidemic to one hive. Any delay which results in the weakening of a colony will invariably mean that stronger stocks will rob out the ailing ones and themselves become infected. Even if drastic robbing does not take place, the diseases can be transmitted by robbing during manipulation or by transferring combs from one hive to another. Furthermore, there is no way of disinfecting hives and appliances against AFB other than subjecting them to scorching, and this is best done by an expert who is accustomed to dealing with the problem as a matter of routine. Even when medicinal treatment is resorted to and the progress of the sickness apparently arrested, the hive and its contents remain infectious and should, both by law and in fairness, not be sold until the bees have been killed and the hive scorched out.

Sacbrood

Another stress disease caused, this time, by a virus, can seriously affect the growth and output of colonies. Sacbrood is so called because the larval skin moults away from the abdomen and thorax, but does not free the head. The pupa thus becomes enclosed in a 'sac' and dies, probably through suffocation. The head quickly darkens to a brown colour as does, more slowly, the body. As the body dries, the head stays high in the cell, giving to the remains the appearance of a 'chinese slipper' with a high raised toe. The dark brown and dried stages of the pupae are not infectious. In bad cases a very large percentage of brood can be affected. The most frequent stress which gives rise to sacbrood comes from damp floor boards during the winter or from situating hives in frost pockets which subject the colony to adverse conditions. The early stages of diseased pupae are highly infectious in the hive but there is less tendency for viral infections to travel from hive to hive. No cure is known, but requeening and improving the over-wintering conditions appears to overcome the disease. During the honey flow the trouble usually ceases on its own. It may well be that an artificial 'flow' in the form of a good feed of syrup could help the colony to overcome the causative stress. Aureomycin appears to protect colonies from infection but does not cure them.

Sacbrood appears to be the only bee disease which can affect both brood and adults. Some adult bees can be found whose normal adult development is upset, with their pollen foraging phase in particular affected. These have been found to be infected with the same sacbrood virus as the brood. It is these infected adults that pass on the virus to the new brood after the winter rest.

Chalk-brood

Like sacbrood, chalk-brood develops in colonies which have suffered some stress, usually caused by chilling the brood at some time during the spring. Cold, exposed sites, damp floors tilted backwards instead of forwards, or without a through passage for air under the hive, or prolonged exposure to air during early spring can allow the fungus *Ascophaera apis* (Maasen) to invade the larva and develop in its lower end. The mycelium of the fungus fills the larva, but leaves the head intact, transforming it into a white chalk-like mummy surmounted by a pale brown head. The black fruiting bodies of the fungus then break through the larva and the spores spread around the hive, and may live there several years. Under good apiary conditions the disease will die out, but it can become a serious pest if the cause of stress is not removed.

Addled and chilled brood

Sometimes a defect in the queen leads to her producing weakened offspring which die in the larval stage. The trouble appears to be genetic and disappears if a new queen is given.

Brood can become chilled during a cold spell in spring when bees have to contract their cluster and leave some of the brood unprotected; the dead brood turns a grey to black colour as it decays before house bees remove the remains. A similar result follows lifting out combs in very cold weather or in biting cold winds.

Adult bee diseases

If brood diseases kill colonies by preventing replacement of adult bees, then fatal sickness of the adults destroys the colony by removing the providers, nurses and others, so that the brood first diminishes and finally starves out. As the bees' lives are short, the pathogens necessarily work fast, and outbreaks can spread very quickly where the natural protection breaks down.

Bee paralysis

A virus infection located mainly in the bees' head and food glands has recently been studied. Severely sick bees tremble and are unable to fly; large numbers of them crawl out of the hive and spread over the ground in front of it. The colonies may die or recover, and it appears likely that most colonies are infected to some degree but only occasionally succomb to the disease. Susceptibility varies from colony to colony, and requeening with a young and healthy queen often produces an apparent cure. The virus is spread by normal food sharing and it does not survive long in dead bees. It appears likely that paralysis causes the death of colonies suffering from other adult diseases and the combination of the two weakens the colonies to the point where they do, in fact, die out. This would explain why nosema, amoeba and acarine disease are often associated with signs of trembling and crawling. As we have no treatment for the paralysis, we have to confine treatment to the other pathogen and this usually, but not always, allows the bees to recover to some extent. Dr L. Bailey of Rothamsted, who is currently studying this disease, considers it to be very widespread and the most severe and quickly fatal disease to which adult bees are subject.

Dysentry

Normally, in winter, bees retain faecal residues in their rectum until they can make a cleansing flight during fine spells. This retention of faeces continues normally until the weight of the faeces reach a third of the bees' total weight. If the weight should rise to 45% the bee must defecate even if it is still in the hive. Abnormal faecal reserves derive from excess water in the winter stores. This excess water can come from unsealed stores of thin syrup fed too late to be processed or from crystallisation of honey in the comb where the crystals become surrounded by dilute fructose. At low dilutions, the sugars ferment and increase the proportion of harmful water. If the whole colony simultaneously suffers from dysentry it may die, and any intestinal pathogen is distributed in the faeces spread over the combs, frames and floor.

Paralysis and dysentry are frequently associated with nosema, amoeba and acarine and the many deaths from these latter diseases may, in fact, be due either to paralysis or to dysentry consequent on bad bee management. Microscopical examination only indicates one of the causes.

Serological tests would be required for detection of paralysis viruses and these are not yet available to beekeepers.

Nosema

Most practical beekeepers will know that the majority of colony deaths in spring occur in colonies infected with the germ *Nosema apis* (Zander) whose oval spores, from infected comb or faeces, are ingested into the mid-gut, where they proliferate and five days later form many spores which either pass into the rectum or reinfect the gut. Infected bees live only half as long as healthy ones so that the colony's gathering potential is considerably reduced, and seriously retards spring development. Under the cool conditions of winter, with the bees confined to the hive, the disease spreads slowly, but dysentry leaves the pathogen spread over the combs in the faeces. As soon as the weather becomes warmer in spring and the brood nest spreads, the cells are cleaned up before eggs are laid in them, and then the microbes spread rapidly among the bees. With the accompaniment of dysentry and paralysis the colony may die, but many survive in a weakened condition. Later in the season, cleansing flights prevent further infection of the comb and the colony may recover. The combs remain contaminated and the following winter will see a recurrence of the trouble, unless new sterile comb is provided during the trouble-free period. Colonies may recover from nosema but usually too late in the season to be productive units.

A number of beekeepers treat their bees annually with 'green' sulphur as a preventive against nosema. The scientists tell us that the treatment is ineffective, but all the practical beekeepers I know who have tried the method assure me that it works. They treat as soon as the honey has been taken off when the bees are still flying freely. A heaped tablespoonful of green sulphur, the finely atomised variety, is put in a $\frac{1}{2}$-lb honey jar and golden syrup slowly mixed in to fill the jar and distribute the sulphur by stir-ring slowly. A tablespoonful of the mixture is placed in a flat tin lid and pushed into the hive entrance. This 'brim-stone and treacle' has its usual purgative effect and, it is claimed, rids the lightly infected bees of the pathogen and kills off the heavily infected ones, so that the infection does not spread in the winter cluster. However, the result-ing bee diarrhoea makes it advisable not to hang washing on the line for a day or two after the medication!

The standard and only really effective treatment against nosema is the administration of syrup medicated with Fumagillin, an antibiotic derived from a common hive fungus and sold under the trade name of Fumidil B. Like most antibiotics it should be kept cool and dry until re-quired. The measured dose, as prescribed on the bottle, should be diluted with warm, but not hot, syrup to make a paste and then stirred well into 4 or 5 litres of syrup and fed to the colony. Opinions vary as to the best time of year for feeding. Some prefer to use Fumidil B as part of the autumn feed, counting on the positive action of the drug to prevent any spread of the disease spores in the winter cluster, others feed it in the spring, feeling that the new season's brood development is better protected. Both points are quite valid and have led me to adopt the prin-ciple of feeding a half dose in the autumn and a further half dose the following spring. In both cases the medicated syrup has to be made up fresh and I merely share the full dose between two stocks. By using Fumidil one can expect a cure rate of between 75% and 100%. But warm damp winters invariably increase the incidence of disease and decrease the percentage of cures in treated colonies.

Although Fumidil B arrests nosema in time to allow the colony a normal development, the combs, the floor and the hive body remain infected and the colony will go through the same cycle again the following winter and spring. A small beekeeper may prefer to rid himself alto-gether of the disease by changing the infected combs for others that are sterile. This can only be done when the spring weather allows the bees to expand the brood, and is only really effective if accompanied by a Fumidil B dose.

Amoeba disease

Malpighamoeba mellificae (Prell), can infect the tubules which serve the bee as a kidney. The general progress of the condition follows so closely the apparent symptoms of nosema that it is only possible to distinguish them under the microscope. Fumidil has no effect on amoeba and infected stocks are usually unproductive for the main flow. But they can be recombed by the Bailey method described below, followed by sterilisation or replacement of the combs, and this sometimes eliminates the disease.

'Bailey' comb change

Put a clean brood chamber with new frames and foundation or old sterilised comb above the brood and add a feeder of syrup—with or without Fumidil. After two or three weeks the queen should be laying in the upper box. When you can see the brood in the new combs, lift off the clean box, which must contain the queen, and place it temporarily over a roof. Then put a queen excluder over the lower brood chamber, put three pieces of 6 to 9 mm thick lath around the back and sides of the excluder and completely block the old lower entrance to the hive. Replace the new box with the queen and brood over the excluder and lath. The queen cannot go below and the bees are now obliged to enter above the excluder. In three weeks time the brood should all be hatched from below, with the exception of some drones, and all the honey stores will have been moved up above. The new box can now be lowered on to a clean floor, not the old one, and the infected box removed for melting down the combs or sterilising. If it still contains many bees, these can be shaken in front of the hive or cleared with a Porter escape overnight into a clean box—but don't leave it longer than necessary to clear the bees. It is most important at this stage to store the infected hive body where other colonies cannot reach it, or to proceed to sterilisation at once. Alternatively, make an artificial swarm and remove the old brood combs for sterilisation three weeks later.

Sterilising comb

The small beekeeper with only two or three hives may prefer to cut out and melt down diseased combs and scrub hives and frames with a stiff brush in a bath of hot water and soda. After refitting the frames with sheets of foundation, they will be clean once again. But good comb is valuable, and in apiaries of any size the cost in cash and time of refitting foundations can be prohibitive. Both combs and hives can be effectively sterilised against nosema or amoeba by maintaining them for 24 hours at a temperature of 55°C in a cabinet, or by fumigating them with acetic acid purchased from a chemist or drug store. It needs to be handled with care as it is a strong acid. If any falls on hands, clothes or metal, wash immediately in running water. My own practice, as soon as a colony dies for any reason, is to close the entrance of the hive; then, on the upper-most box I put a saucer or foil dish containing a piece of wadding and pour on about 15 ml (a tablespoonful) of acetic acid, then an empty super, covered with a sheet of newspaper and a roof. The hive is then left for a period of five to fifteen days, after which the wadding and acid should be removed. The boxes are then skewed at 45° to one another to allow a draught to dissipate the fumes for a few days to free them of poisonous vapours. It must be stressed that AFB and EFB germs are not destroyed by this method. The fumigation should not last more than a week to avoid corrosion of the frame nails and wire. In this way all the contents of an infected hive are effectively fumigated and made safe.

Acarine or 'Isle of Wight' disease

It appears likely that the famous epidemic of Isle of Wight disease which decimated the bee population of the British Isles from 1905 to 1924 was a combination of Acarine infestation and bee paralysis. Acarine is so called from a mite *Acarapis Woodii* (Rennie) which penetrates the breathing tubes or tracheae in the thorax of quite young bees and proliferates there. If paralysis and dysentry are also present, and they usually are, the combs become soiled with faeces which spread the paralysis virus. The bees show all the symptoms of severe paralysis and the acarine mites can be found in their thoracic tracheae. If medication, which kills the mites, is applied to the colony the bees seem able to overcome the other conditions.

Two remedies are commonly recommended for acarine: Frow mixture and Folbex fumigation. The original Frow mixture consisted of safrol, ligroin or petrol and nitrobenzene, but it was later found that only the nitrobenzene was effective. It can only be applied in the months when no brood is present in the hive, because the fumes kill brood. A shallow tin or foil pan, holding a piece of cotton wool about 7 cm square, is dosed with a teaspoonful of Frow or nitrobenzene and placed above the winter cluster. The treatment should be repeated three times at five day intervals, preferably during a really cold winter spell.

Folbex treatment can be applied at any time of the year, even when the bees are flying freely, but to be effective has to be applied in the evenings when all flying bees have returned home. The active ingredient is soaked into blotting paper with some potassium chlorate and sold in strips, each sufficient for one fumigation. A tin lid or foil saucer serves as a base. The strip is folded down the middle into a long tent shape, placed on the tin, ignited and pushed into the hive entrance. A sheet of paper is quickly pinned over the entrance to prevent egress of bees. The bees will chew away an exit as soon as they can. A foam plastic strip can be used to block the entrance but should be removed from 1 to 2 hours later. Good results can be obtained by treating three times at five-day intervals, but a complete cure can seldom be achieved with fewer than five treatments.

Combs of bees that have died out through acarine and its complications are not infective after the death of the colony, and paralysis virus disappears after a few days.

Expert diagnosis

In many countries there exist free laboratory services for the diagnosis of bee diseases and chemical poisoning. A list is appended for the use of readers from England and Wales. Others would be well advised to obtain the relevant addresses from their National Beekeeping Association well before they are needed. A few rules for sending samples should be followed to ensure their safe arrival in a condition which permits meaningful results to be obtained by the laboratory technicians.

Containers made of air-tight materials such as glass, plastic or metal should never be used for samples either of brood or adult bees; instead use small card boxes, such as matchboxes, and wrap them in paper. Do not include honey, syrup or candy for the bees to feed on. It will only kill them more quickly, and make them sticky and unpleasant to handle in the laboratory. For brood diseases call in the bees' officer whenever possible, but if this is not possible, cut out a piece of comb about 10 cm × 15 cm so as to include the suspected cells, wrap it in several thicknesses of newspaper, lay this in a cardboard box with your name and address, add query and hive number in an envelope, and finally wrap the box in firm brown paper. Address it to your diagnostic station, which for England and Wales is:

The Bee Disease Specialist,
ADAS,
Trawscoed,
Near Aberystwyth,
Dyfed.

Scraps of comb and the paper you cleaned the knife with should be burned at once.

When adult bee disease is suspected, a sample of thirty live or moribund bees should be collected in a matchbox and either sent to the same address or to the nearest county diagnosis centre. To take a sample of bees use an empty matchbox covered with a small piece of glass. By lifting out a comb and tilting the matchbox tray, a populous area can be swept to imprison about thirty bees. If the tray is then slid over the glass and upturned the actual number of bees can be counted through the glass. Should the total be less than thirty, slide the tray into the box away from under the glass and further additions can be made by inserting the bees one by one. Bees can be picked up by holding them by both wings. They cannot then twist their bodies to sting. Mark on the box the hive number or position. Enclose a note with your name, address and query, and wrap the box or boxes in several layers of stiff brown paper.

When you suspect poisoning, the procedure is somewhat more complex, because the methods of analysis demand a lengthy process of elimination among the vast number of compounds present in minute amounts. It is not reasonable to expect the laboratory to apply tests for the whole range of poisonous substances involved. The diagnosis can only be successful if some guidance is sent with the sample, giving clues to the possible chemical involved. The beekeeper may only become aware of the damage when large numbers of bees lie dead or dying in front of his hives. A sample of about one quarter of a litre should be gathered into a cardboard box, and as much information as possible included in the accompanying letter. Particularly relevant are:

(a)   The extent of the damage, number of hives affected and how severely.

(b)   The type of crop growing nearby which has been sprayed or dusted.

(c)   The grower or spray contractor, with names and addresses.

(d)   The nature or trade name of the substance used, when this can be obtained.

For beekeepers in England and Wales, the parcel should be sent to:

> The Beekeeping Advisory Officer,
> ADAS,
> Rothamsted Lodge,
> Hatching Green,
> Harpenden,
> Herts.

At the same time it is advisable to notify the farmer or grower that you are sending in these samples and that if the tests, which take some weeks, prove positive, that you hold him responsible for the damage and that you will be claiming compensation. It is at such times that membership of a beekeepers' association proves to be a great help.

Bee 'lice'

Infestations of a small wingless fly, *Braula coeca* (Nitz), seem to cause little damage to the bees but certainly spoil the cappings of comb honey. This little pest clings to the thorax of the bees; workers, queen and occasionally even drones. When the flies are hungry they move onto the bee's head and obtain food by stroking the antennae and mouth parts of their host, who is decieved into offering food as to a soliciting worker. The queen is particularly subject to infestation and as many as a hundred can cling to her body, impeding her movements and interfering with pheromone distribution. Once they have fed, the braulae return to their waiting position on the bee's thorax where they cling with the greatest tenacity. It is difficult to remove them even with tweezers. They lay eggs in the cell cappings, where their minute larvae build tunnels. Sections that have been attacked in this way are virtually unsaleable, and anyone wishing to produce comb honey, either in sections or in the form of cut comb, should take measures to eliminate braulae from his hives.

Cures are uncertain; all rely on some form of fumigation and, as usual with parasites, only the adults are affected by the fumes. Repeated treatment is required to kill the new adults which subsequently hatch out before they have time to breed again. In practice three weekly treatments prove to be effective. It must be noted that adult braulae are not killed by the fumigation, but merely drop temporarily insensible onto the floor of the hive, where they recover and attach themselves to new hosts as they come in and out of the hive. The solution is to insert a piece of stiff paper or thin card large enough to cover the inside of the tray, then to fumigate and remove the paper and burn the parasites. In the past tobacco fumigation through the smoker was usually recommended and proved effective, but research has shown that tobacco smoke affects the subsequent performance of the queen by prematurely ageing her. The Frow treatment for Acarine is useful in reducing a heavy infestation in the winter, but the most up-to-date cure is a fumigation with a proprietary chemical called 'Phenovis', obtainable from most veterinary chemists either in the form of a powder or in 1 gram tablets. The smoker is lit and operated until a good hot base is burning in the fire box, and 3 gm of powder or crushed tablets in a loose paper is inserted into the smoker. The fumes are then blown into the hive in the evening or early morning. The parasites will fall onto the paper in about three minutes and can be removed and destroyed. Three treatments at least should be given at weekly intervals. Since braula is an insect it is very difficult to find a substance powerful enough to kill them and not harm our bees at the same time.

Wax moths

Two moths have evolved systems for digesting wax as the carbohydrate part of their diet. In order to reach their full development they do need the additional protein which they can find in the cast-off skins of the bee larvae or residues of pollen. The lesser wax moth, *achroia grisella*, breeds in cooler climates than the larger *galleria mellonella*, which can be a very serious pest indeed in the warmer regions of the world. Here in England, sound apiary hygiene and sensible modern hives allow the bees to cope adequately with normal attacks. Further control can be obtained by the use of a queen excluder which prevents breeding or excessive pollen storage in the supers. The moth larvae, which hatch on clean wax comb, do not survive their first moult and very little damage indeed results. In common practice, stored combs which contain neither pollen nor cast-off larval skins are quite unable to sustain growth and development of the wax moth larvae. The moral here is obvious. Super comb, normally taken off in winter and stored, is safe from serious moth damage, because the queen excluder prevents the queen from laying and the workers from storing pollen in those combs. If disinfected or sterilised brood combs are kept, the wax moth damage will be small during hard cold winter weather; the danger period is in the autumn and spring, when the temperature permits the eggs to hatch. Damage can be prevented by enclosing such combs, in their hive box, in large plastic bags with a tablespoonful of PDB (Paradichlor-benzine) crystals. These give off an insecticidal gas, so great caution must be exercised. Before combs treated in this way can be used again, they must be 'aired' for about a week to rid them of any trace of the killer gas. In practice it is probably better to melt down any comb that has been bred in, rather than try to sterilise and store it.

In warmer climates the large wax moth breeds sufficiently rapidly to damage and destroy the combs in well-populated colonies. Once the moth larvae have hatched they bore through the brood combs, parallel to the midrib, lining their tunnels with extremely tough silk. Whenever these tunnels are seen on the brood comb, push a corner of the hive tool to either end of the run and the moth larvae can then be pulled out and destroyed. The moth often tunnels just under the brood cappings, consuming the caps on their way. The bee larvae continue to develop without the usual covers. This 'bald brood' shows that the wax moth is active and a check should be made to see if there are any areas in the hive where the moth larvae can hide away from the bees. The types of hive recommended in this book will reduce the likelihood of serious moth attacks. But the type of frame with 'metal ends' and the cloth, canvas or carpet quilts that are frequently found in some older apiaries, provide the ideal breeding areas and conditions for these and other pests. Eventually they pupate in extremely tough cocoons, until they emerge and disperse to lay further eggs. A new treatment with a micro-organism is being developed which destroys the wax larvae and does little harm to the bee larvae. This may be of great value in some sub-tropical areas in controlling *galleria*.

Although bee diseases are prevalent, normal common sense cleanliness avoids much damage from infestation. There is certainly no need to be put off beekeeping because of disease problems, and in most countries the authorities usually provide diagnostic services and curative advice, or even free treatment. If, as a result of some disaster, your bees do die of a disease, then block completely the entrance to the hive as soon as possible, to safeguard other hives in the neighbourhood from contracting the same sickness by robbing out the 'dead' hive.

I have found that the 'cures' mentioned above have only limited application. Bees resistant to most diseases do exist. The survivors in an apiary decimated by disease are usually resistant. If the cure can be applied as a temporary measure to tide the colony over until it can be requeened from a more resistant strain, then surely this must be better than constant treatment with expensive biochemicals.

The emphasis in beekeeping is always on health; bees are normally healthy creatures, hive products are clean and health giving and beekeepers can rest assured that their craft will promote their own health, and happiness.

Appendix 1: Useful Addresses

For all information and world wide contacts:

The Bee Research Association
Hill House, Chalfont St Peter, Gerrards Cross, Bucks.,
England.

The BRA also has a bee library, picture and slide library, translations, photocopies and computer printouts of beekeeping bibliographies, and a collection of Historic and Contemporary Beekeeping Material. It publishes *Bee World*, *Journal of Apicultural Research* and *Apicultural Abstracts*.

For diagnosis of adult bee diseases and spray poisoning:

The Bee Advisory Officer,
ADAS Rothamsted Lodge, Hatching Green, Harpenden,
Herts.

For diagnosis of brood diseases:

The Bee Expert,
ADAS Trawscoed, Nr. Aberystwyth, Dyfed.

The present address of the British Beekeepers Association is:

The General Secretary, BBKA
55, Chipstead Lane, Riverhead, Sevenoaks, Kent.

Appliance Manufacturers:

Robert Lee, Beehive Works, George Street, Uxbridge,
Middlesex.

Steele and Brodie, Beehive Works, Wormit, Fife, Scotland.

E. H. Taylor Ltd., Beehive Works, Welwyn, Herts.

E. H. Thorne (Beehives) Ltd., Beehive Works, Wragby,
Lincolnshire LN3 5LA.

Appendix 2: Reference and Suggested Reading

ABC and XYZ of Bee Culture, A. I. Root (Root)
Anatomy and Dissection of the Honeybee, H. A. Dade (BRA)
Beekeeping, Eckert & Shaw (Macmillan)
Behaviour and Social Life of Honeybees, R. Ribbands (BRA)
Honeybees from Close-up, A. M. Dines (Cassell)
Infectious Diseases of the Honeybee, L. Bailey (Land Books)
Introduction of Queen Bees, L. E. Snelgrove (Snelgrove)
Pollination of Fruit Crops, HEA (Elvey & Gibbs)
Queen Rearing, Laidlaw and Eckert (University of California)
Swarming. Its Prevention and Control, L. E. Snelgrove (Snelgrove)
World of the Honeybee, C. G. Butler (Collins)

Ministry of Agriculture Bulletins

No.			
9	Beekeeping	144	Beehives
100	Diseases of Bees	206	Swarming of Bees
134	Honey from Hive to Market		

Ministry of Agriculture free advisory leaflets

No.		No.	
306	Foul Brood	412	Feeding Bees
328	Importance of Bees in Orchards	468	Modified Commercial Hive
330	Acarine Disease	473	Nosema and Amoeba Disease
344	Migratory Beekeeping	549	Langstroth and Modified Dadant Hives
347	Beeswax from the Apiary		
362	Examination of Bees for Acarine	561	Minor Brood Diseases
367	British National Hive		

Appendix 3: Metric Conversion Tables

Metres/Inches

mm	inch	mm	inch	mm	inch
1	$\frac{1}{25}$	43	$1\frac{11}{16}$	260	$10\frac{1}{4}$
2	$\frac{1}{12}$	48	$1\frac{9}{20}$	286	$11\frac{1}{4}$
3	$\frac{1}{8}$	51	2	292	$11\frac{1}{2}$
5	$\frac{3}{16}$	76	3	298	$11\frac{3}{4}$
6	$\frac{1}{4}$	108	$4\frac{1}{4}$	305	12
8	$\frac{5}{16}$	114	$4\frac{1}{2}$	356	14
9	$\frac{3}{8}$	121	$4\frac{3}{4}$	413	$16\frac{1}{4}$
12·5	$\frac{1}{2}$	140	$5\frac{1}{2}$	419	$16\frac{1}{2}$
16	$\frac{5}{8}$	146	$5\frac{3}{4}$	431	17
18	$\frac{3}{4}$	152	6	448	$17\frac{5}{8}$
22	$\frac{7}{8}$	159	$6\frac{1}{4}$	460	$18\frac{1}{8}$
25	1	216	$8\frac{1}{4}$	463	$18\frac{1}{4}$
27	$1\frac{1}{16}$	223	$8\frac{3}{4}$	508	20
35	$1\frac{3}{8}$	232	$9\frac{1}{8}$	546	$21\frac{1}{2}$
37	$1\frac{9}{20}$	239	$9\frac{3}{8}$	552	$21\frac{3}{4}$
38	$1\frac{1}{2}$	243	$9\frac{9}{16}$	559	22
42	$1\frac{5}{8}$	254	10		

Centigrade/Fahrenheit

Cent.	Fahr.	Cent.	Fahr.
0	32	49	120
5	40	54	130
7	44	60	140
30	86	62	144
34	92	82	180
38	100	90	194
43	110	100	212

Appendix 4: Beekeeping

During the inevitable lapse which occurs between the writing and publication of any book, certain changes have taken or are due to take place and some discoveries have been made public.

The British Beekeepers' Association has decided in future to use the more euphemistic terms: 'American brood disease' and 'European brood disease' instead of American or European foulbrood.

The Bee Research Association may, before publication date, already have changed its name to International Bee Research Association; thus reflecting in its title the true spread of its sphere of influence.

The EEC standards for bottled honey for sale will be issued currently with the publication of this book. Its provisions will concern the honey packing industry rather than the small beekeeper. However, packs of cut comb for sale will have to be labelled with the net weight of the honeycomb. This has not hitherto been required in England.

An interesting piece of scientific research has just been published which lends some substance to my guess, on page 54, that electrostatic generated by nylon type fabrics stimulate bees to stinging. It would appear that bees in hives situated close to high-tension cables sting more and show noticeably more aggressive behaviour than when they are located beyond the electrical field which surrounds the wires. Note should be taken of this when siting apiaries (Chapter 5).

Index